Principles and Applications of Vector Network Analyzer Calibration Techniques

RIVER PUBLISHERS SERIES IN COMMUNICATIONS AND NETWORKING

The "River Publishers Series in Communications and Networking" is a series of comprehensive academic and professional books which focus on communication and network systems. Topics range from the theory and use of systems involving all terminals, computers, and information processors to wired and wireless networks and network layouts, protocols, architectures, and implementations. Also covered are developments stemming from new market demands in systems, products, and technologies such as personal communications services, multimedia systems, enterprise networks, and optical communications.

The series includes research monographs, edited volumes, handbooks and textbooks, providing professionals, researchers, educators, and advanced students in the field with an invaluable insight into the latest research and developments.

Topics included in this series include:-

- Communication theory
- Multimedia systems
- Network architecture
- Optical communications
- Personal communication services
- Telecoms networks
- Wifi network protocols

For a list of other books in this series, visit www.riverpublishers.com

Principles and Applications of Vector Network Analyzer Calibration Techniques

J. Apolinar Reynoso Hernández

Centro de Investigación Científica y de Educación Superior de Ensenada, Baja California, México

Manuel Alejandro Pulido Gaytan

Centro de Investigación Científica y de Educación Superior de Ensenada, Baja California, México

Published 2024 by River Publishers
River Publishers
Alsbjergvej 10, 9260 Gistrup, Denmark
www.riverpublishers.com

Distributed exclusively by Routledge
605 Third Avenue, New York, NY 10017, USA
4 Park Square, Milton Park, Abingdon, Oxon OX14 4RN

Principles and Applications of Vector Network Analyzer Calibration Techniques / by J. Apolinar Reynoso Hernández, Manuel Alejandro Pulido Gaytan.

Routledge is an imprint of the Taylor & Francis Group, an informa business

ISBN 978-87-7004-159-1 (hardback)
ISBN 978-87-7004-651-0 (paperback)
ISBN 978-87-7004-644-2 (online)
ISBN 978-87-7004-635-0 (master ebook)

To my wife, Fabiola, and my daughter, Myriam.
J.A. Reynoso-Hernandez

To my family.
Manuel Pulido

Contents

Foreword

"Prof. Reynoso Hernandez is leading the premier microwave measurement laboratory in Mexico and Latin America. His RF and Microwave Laboratory, one of the best-equipped labs in Latin America, has been the origin of many seminal progresses and achievements in the field of measurements. His lab's expertise encompasses both linear and nonlinear microwave measurements.

My first acquaintance with Prof. Reynoso Hernandez took place at the Automatic Radio Frequency Technical Group conference when I noticed that his group was winning several years in a row best paper prizes in the field of calibration theory. I had the opportunity to visit his laboratory in Ensenada, Baja de California. Using the latest technology, some of the most talented students in Mexico and Latin American countries have been working there. Most certainly, the breath taken view of the Pacific coast from his lab has inspired their work.

I had the chance to co-advise a couple of his PhD graduate students: Dr. Andres Zarate and Thaimí Niubó. Dr. Zarate demonstrated the application of ANN to the modeling of SOI MOSFETs. Dr. Andrés Zarate is leading the PSemi laboratory in San Diego, CA. Among her achievements, Dr. Thaimí Niubó demonstrated the first calibrated measurement of broadband constellations with a VNA. This feature is now commercialized by Keysight. Dr. Thaimi Niubo Aleman is now working with SMK Electronics.

His students have all pursued distinguished carriers. Dr. Manuel Pulido-Gaytan, the book co-author completed his PhD with Prof. Reynoso Hernández in 2016. Jointly with Prof. Reynoso-Hernández, he has made significant contributions to the state of the art of VNA calibration theory and load-pull measurement systems. He has several years of experience working on the characterization and modeling of RF and microwave devices in several organizations within the US semiconductor industry. He is now working at Skyworks Solutions Inc., in Irvine, CA, where he works as Principal Engineer of the Device Modeling Group.

Lately, Prof. Reynoso Hernández has been my "companion d'arme" in running the NVNA Users' Group at ARFTG and IMS. Prof. Reynoso Hernández is currently an associate editor for the IEEE Transactions on Microwave Theory and Techniques.

Prof. Reynoso Hernández did his PhD at the University of Toulouse, France under the direction of Prof. Jacques Graffeuil, a world recognized expert in the area of MESFET's, HEMT's, HBT's noise and nonlinear device modeling and characterization, MMIC design and in the characterization of traps by using Low frequency noise measurements.

His microwave measurement expertise is thus running for devices to circuits. Recently his lab developed a unique low-frequency NVNA testbed which he used to verify the practical realization in different semiconductor technologies of the novel continuous class-J and class-F at the current source reference planes."

Prof. Patrick Roblin
Ohio State University

Preface

Vector network analyzers (VNAs) are instruments that are used to measure the scattering parameters of microwave devices, which in turn may be used to develop linear and nonlinear models of these devices. In order to remove the effect of hardware non-idealities, the VNA must be calibrated. These non-idealities are modeled by using an error model. The 8-term error model and 12-term error model are often used to account for non-idealities in two-port VNA hardware. Then during the VNA calibration process, the first step is to identify the error model to be used, and then utilize a set of calibration elements to estimate these error terms. The names of the different calibration techniques derive from the type of utilized calibration elements. In this sense, several calibration techniques have been developed and widely used in the industry and academy.

The first calibration technique used to calibrate the VNA was the Short-Open-Load-Thru (SOLT). This calibration technique uses the 12-term error model. The main characteristic of SOLT calibration technique is the use of the calibration standards (known electrical behavior) instead of calibration structures (unknown or partially known electrical behavior). On the other hand, the calibration techniques using the eight-term error model are: thru-reflect-line/line-reflect-line (TRL/LRL), thru-reflect-match/line-reflect-match (TRM/LRM), and thru-reflect-reflect-match/line-reflect-reflect-match (TRRM/LRRM). Another technique using the eight-term error model is the line-offset open-offset short (LZZ), which is a contribution of the authors of this book to the state-of-the-art VNA calibration techniques.

In the implementation of the calibration techniques, transfer parameters (T-parameters), also known as wave cascade matrix, and chain matrix (ABCD-parameters) are very often used to identify the error model terms. This book uses both of these approaches to present mathematical descriptions of different VNA calibration techniques.

A crucial parameter in all the calibration techniques is the reference impedance. In this book all the calibration techniques are developed under the philosophy that the only reference impedance to which the calibration should be referred is the measurement system impedance. Every calibration technique is developed from the basics, up to an advanced level, so that the reader may be self-guided to a desired level of expertise.

This book summarizes the work developed during more than two decades in the field of *Advanced Calibration Techniques for Vector Network Analyzers*, by the *RF and Microwave Group* at *The Center for Scientific Research and Higher Education of Ensenada, Baja California, Mexico*, which is led by Dr. J. Apolinar Reynoso-Hernandez, author of this book.

Vector network analyzers are normally used by engineers and researchers working on the RF and microwave field, which usually requires advanced and specialized courses at graduate level. This book is written so that every electrical engineer, with knowledge of electrical circuits and linear algebra basics, is able to understand the principles of VNA calibration techniques. The reader should be able to implement any VNA calibration technique, decide the most adequate calibration for a given measurement condition, and interpret the measurement results, as a seasoned RF metrology expert.

Chapter 1 of this book is focused on providing an introduction to the concepts the rest of the book is devoted to. The definition and measurement of scattering parameters are presented. Then, a definition of a large-signal measurement system is introduced, so that the reader can understand the extension of small-signal measurement systems into systems able to characterize devices in their large-signal regime.

In Chapter 2, the different error models used to describe the imperfections of a vector network analyzer are presented. The one-port error model (3 terms) is used as a foundation for the more complex two-port (12 term) error model. Then, the switching error correction is presented as a means to further simplify the 12-term error model into an 8-term error model.

Chapter 3 of this book deals with the concept of microwave de-embedding, in which the classical de-embedding method for devices embedded in test fixtures is described in detail. Then, the straightforward de-embedding method is presented. These two methods set the basis for the further introduction of calibration techniques for vector network analyzers.

The principles presented in Chapters 2 and 3 are used as the basis for the VNA calibration techniques presented in Chapter 4. Chapter 4 presents what may be called "The ABCD-parameters of VNA calibration techniques". The ABCD-parameters matrix formalism is used to describe the electrical behavior of a device under test and the vector network analyzer. Then, VNA calibration algorithms that are developed using this matrix formalism are presented.

In Chapter 5, several methods for characterizing transmission lines are presented. The first method is the L-L method in which it is demonstrated that the propagation constant and characteristic impedance of a transmission line may be calculated by using two lines of different lengths but identical impedance and propagation constant. The second method to be described is the calibration comparison method, in which the calibration terms obtained from the measurement of the calibration structures corresponding to two different calibration techniques are used to determine the characteristic impedance of a transmission line. The third method is a novel procedure that demonstrates that the characteristic impedance of a line may be determined without the need to calibrate a vector network analyzer.

Chapter 6 presents the theory, implementation, and calibration of vector-receiver load-pull systems. The theory reviewed in previous chapters is used to develop calibration algorithms for the calibration of large-signal measurement systems.

J. Apolinar Reynoso-Hernandez
Manuel A. Pulido-Gaytan

Acknowledgements

There are many people the authors would like to acknowledge for being an important part in the work behind this book. The authors hope they did not miss anyone, and otherwise would like to sincerely apologize. Authors would like to acknowledge all the current and former members of CICESE RF and Microwave Research Group, whose work has directly or indirectly impacted the research on VNA calibration techniques. The technical staff of CICESE Research Center are acknowledged, whose work has been imperative for the success of this work: Jesus Ibarra Villasenor, Benjamin Ramirez Duran, and Rene Torres Lira. Thanks to internal and external collaborators, whose work and discussions have contributed to enrich the work behind this book: Maria del Carmen Maya Sanchez, Jose Raul Loo, Roberto Murphy, Carmen Fabiola Estrada Maldonado, Mariano Aceves Mijares, Ricardo Cuesta, and Ignacio Zaldivar. A special acknowledgement to all the MSc and PhD students whose work is related to the development, or improvement, of VNA calibration techniques and practices, specially to those whose thesis works were directly reflected in this book.

Finally, the authors would like to thank the Mexican Council for Science and Technology (CONACyT) and the Center for Scientific Research and Higher Education of Ensenada, Baja California.

List of Figures

List of Tables

List of Abbreviations

DC	Direct current
DUT	Device under test
ECM	Equivalent circuit model
FET	Field-effect transistor
HFET	Hetrostructure Field Effect Transistor
L-M	Line-match
LP	Load-pull
LRL	Line-reflect-line
LRM	Line-reflect-match
LRRM	Line-reflect-reflect-match
LZZ	Line-offset open-offset short
LZZM	Line, open, short, unknown load
mTRL	Multi-line TRL
NVNA	Nonlinear vector network analyzer
PA	Power amplifier
PAE	Power added efficiency
SOLT	Short-Open-Load-Thru
T-A	Thru-attenuator
TRL	Thru-reflect-line
TRM	Thru-reflect-match
TRRM	Thru-reflect-reflect-match
VNA	Vector network analyzer
VR	Vector-receiver
VRLP	Vector-receiver load-pull
WCM	Wave cascading matrix

1

Fundamentals

1.1 Introduction

Microwave network analysis is the formalism by which the electrical performance of microwave circuits is analyzed. It is a method for characterizing microwave circuits by evaluating their response in amplitude and phase to a test signal. Measuring both the magnitude and phase of signals coming to and emerging from components is important for a number of applications in RF and microwave research; some of these applications are device characterization, circuit design, device modeling, and calibration.

In this chapter, the fundamentals of vector network analysis will be reviewed. The discussion includes the common parameters that can be measured at high frequencies, such as S-parameters, and other important parameters that are useful for the readership in the following chapters, such as T-parameters and ABCD-parameters. Finally, a summary of the content of the chapters included in the remaining of the book is also provided.

1.2 Scattering Parameters

Since more than half a century, S-parameters [47] have been one of the most important foundations of microwave theory and techniques. S-parameters are easy to measure at high frequencies with a VNA, provided that systematic measurement errors have been removed by using appropriate calibration procedures. S-parameters allow to estimate in a straightforward manner important characteristics of a device under test (DUT), such as small-signal gain, reflection coefficients, delay, etc.

The term *scattering* refers to the relationship between incident and scattered (reflected or transmitted) traveling waves. We can distinguish between an incident wave a and a scattered wave b [69]. These wave variables are defined as simple linear combinations of the voltage (v) and current (i) flowing

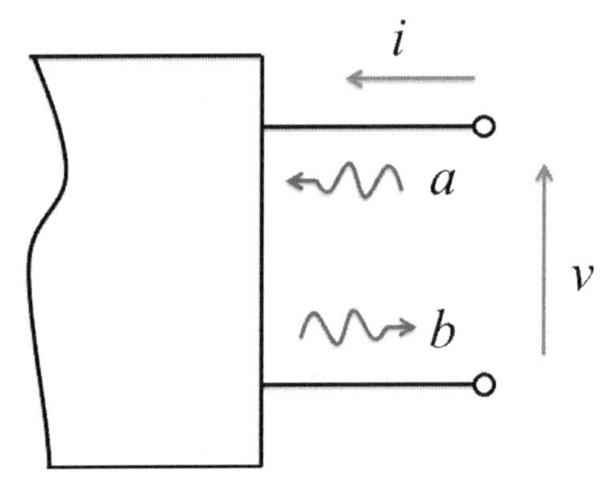

Figure 1.1 Wave definitions.

through the same port (Figure 1.1) according to the following expressions:

$$a = \frac{v + iZ_0}{2\sqrt{Z_0}}, \tag{1.1}$$

$$b = \frac{v - iZ_0}{2\sqrt{Z_0}}. \tag{1.2}$$

The reference impedance for a given port, Z_0, is in general a complex quantity. Nevertheless, for most practical cases in microwave engineering, it is desired that Z_0 be purely real and frequency independent. This concept is of paramount importance in microwave systems' calibration.

Analogously, by solving eqn (1.1) and (1.2) for v and i, the current and voltage can be recovered from the wave parameters as follows:

$$v = \sqrt{Z_0}\,(a + b)\,, \tag{1.3}$$

$$i = \frac{1}{\sqrt{Z_0}}\,(a - b)\,. \tag{1.4}$$

The variables in eqn (1.3) and (1.4) are complex numbers representing the RMS phasor description of sinusoidal signals in the frequency domain. Thus, *a, b, v,* and *i* are vectors, the components of which indicate the values associated with sinusoidal signals at a particular port i, where $i = 1, 2, ..., N$, with *N* defined as the total number of ports in the network. Figure 1.2 shows a graphical representation of the description of these waves for the case of a two-port network.

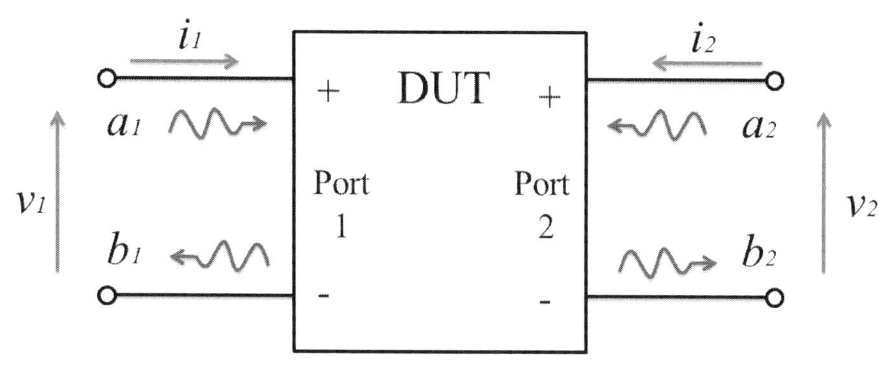

Figure 1.2 Incident and scattered waves of a two-port network.

The assumption behind the S-parameters formalism is that the system being described is linear and therefore there is a linear relationship between the phasor representation of incident and scattered waves, as expressed below

$$b_i = \sum_{j=1}^{N} S_{ij} a_j, \forall i \in 1, 2, ..., N. \tag{1.5}$$

The set of complex coefficients S_{ij} in eqn (1.5) defines the S-parameters of an N-port device. The summation in eqn (1.5) is over all port indices, so that incident waves at each port j contribute in general to the overall scattered wave at each output port i. For the particular case of a two-port network, the scattered waves at ports 1 and 2 of the network are expressed as

$$b_1 = S_{11} a_1 + S_{12} a_2, \tag{1.6}$$

$$b_2 = S_{21} a_1 + S_{22} a_2. \tag{1.7}$$

S-parameters can be grouped to obtain the S-parameters matrix as follows:

$$\begin{bmatrix} b_1 \\ b_2 \end{bmatrix} = \begin{bmatrix} S_{11} & S_{12} \\ S_{21} & S_{22} \end{bmatrix} \begin{bmatrix} a_1 \\ a_2 \end{bmatrix}. \tag{1.8}$$

Thus far, the concept of S-parameters as a linear combination of incident and scattered waves has been presented. In the following sections, other circuit representations commonly used in microwave circuit analysis (ABCD-parameters and T-parameters), which are also used in the development of VNA calibration techniques, are presented.

1.3 Wave-cascading Transmission Parameters

T-parameters are a modified version of S-parameters, which facilitates the representation of the cascading connection of two or more networks, as outlined next. By rearranging the terms in eqn (1.6) and (1.7), such that the variables at the left-hand side of the equal sign are the waves at port 1, a_1 and b_1, and the variables at the right-hand side of the equal sign are the waves at port 2, a_2 and b_2, one has

$$\begin{bmatrix} b_1 \\ a_1 \end{bmatrix} = \begin{bmatrix} T_{11} & T_{12} \\ T_{21} & T_{22} \end{bmatrix} \begin{bmatrix} a_2 \\ b_2 \end{bmatrix}, \tag{1.9}$$

where

$$\begin{bmatrix} T_{11} & T_{12} \\ T_{21} & T_{22} \end{bmatrix} = \frac{1}{S_{21}} \begin{bmatrix} -\det(S) & S_{11} \\ -S_{22} & 1 \end{bmatrix}. \tag{1.10}$$

Thus, for a two-port network, such as the one depicted in Figure 1.3, the T-parameters are represented as a function of incident and reflected waves as

$$T_{11} = \left.\frac{b_1}{a_2}\right|_{b_2=0}, \tag{1.11}$$

$$T_{12} = \left.\frac{b_1}{b_2}\right|_{a_2=0}, \tag{1.12}$$

$$T_{21} = \left.\frac{a_1}{a_2}\right|_{b_2=0}, \tag{1.13}$$

$$T_{22} = \left.\frac{a_1}{b_2}\right|_{a_2=0}. \tag{1.14}$$

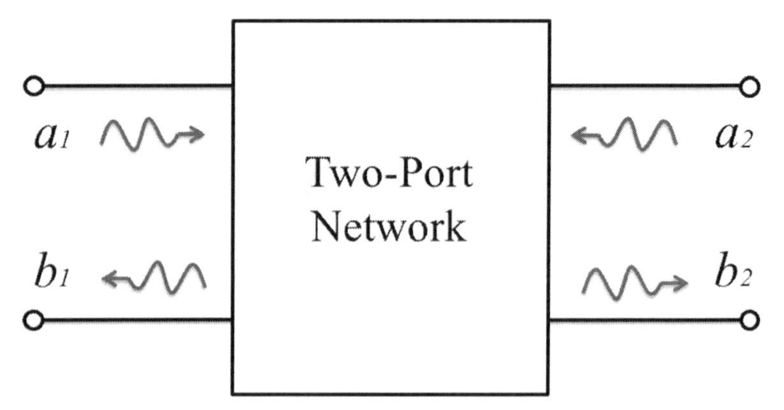

Figure 1.3 Two-port network and the incident and reflected waves at its terminals.

The T-parameters network representation allow representing the cascading connection of N networks by simple multiplication of their individual T-parameters matrices as

$$\begin{bmatrix} b_{1T} \\ a_{1T} \end{bmatrix} = \begin{bmatrix} T1_{11} & T1_{12} \\ T1_{21} & T1_{22} \end{bmatrix} \begin{bmatrix} T2_{11} & T2_{12} \\ T2_{21} & T2_{22} \end{bmatrix} \cdots \begin{bmatrix} TN_{11} & TN_{12} \\ TN_{21} & TN_{22} \end{bmatrix} \begin{bmatrix} b_{2T} \\ a_{2T} \end{bmatrix}, \tag{1.15}$$

where a_{iT} and b_{iT}, $(i = 1, 2)$, represent the incident and reflected waves at the input and output ports of the overall structure.

1.4 Voltage-Current Transmission Parameters

ABCD-parameters, also referred to as chain parameters, are the analogy of T-parameters for voltages and currents instead of incident and scattered waves. The ABCD-parameters representation of a two-port network allows relating the voltage at and the current flowing through port 1 with the corresponding voltages and currents existing at port 2 of a two-port network, as shown in Figure 1.4. The following equation represents what is known as the ABCD-parameters matrix of a two-port network:

$$\begin{bmatrix} v_1 \\ i_1 \end{bmatrix} = \begin{bmatrix} A & B \\ C & D \end{bmatrix} \begin{bmatrix} v_2 \\ i_2 \end{bmatrix}, \tag{1.16}$$

where

$$\begin{bmatrix} A & B \\ C & D \end{bmatrix} = \begin{bmatrix} (1+S_{11})(1-S_{22})+S_{12}S_{21} & Z_0\left((1+S_{11})(1+S_{22})-S_{12}S_{21}\right) \\ \frac{1}{Z_0}\left((1-S_{11})(1-S_{22})-S_{12}S_{21}\right) & (1-S_{11})(1+S_{22})+S_{12}S_{21} \end{bmatrix}. \tag{1.17}$$

The four parameters of this matrix can be expressed as a function of the voltages at and currents flowing through ports 1 and 2, under open circuit (zero current) and short circuit (zero voltage) conditions at port 2:

$$A = \left.\frac{v_1}{v_2}\right|_{i_2=0}, \tag{1.18}$$

$$B = \left.\frac{v_1}{i_2}\right|_{v_2=0}, \tag{1.19}$$

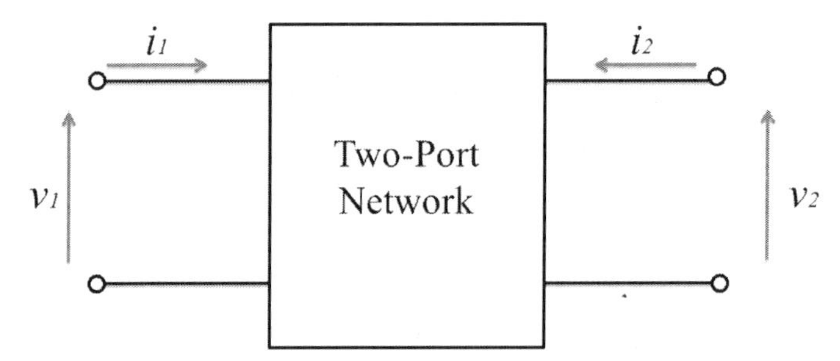

Figure 1.4 Two-port network and the voltage and current at its terminals.

$$C = \left. \frac{i_1}{v_2} \right|_{i_2=0}, \tag{1.20}$$

$$D = \left. \frac{i_1}{i_2} \right|_{v_2=0}. \tag{1.21}$$

As in the case of T-parameters, the ABCD-parameters allow representing the cascading connection of N networks by multiplying their individual ABCD-parameters matrix representation as

$$\begin{bmatrix} V_{1T} \\ I_{1T} \end{bmatrix} = \begin{bmatrix} A_1 & B_1 \\ C_1 & D_1 \end{bmatrix} \begin{bmatrix} A_2 & B_2 \\ C_2 & D_2 \end{bmatrix} \cdots \begin{bmatrix} A_N & B_N \\ C_N & D_N \end{bmatrix} \begin{bmatrix} V_{2T} \\ I_{2T} \end{bmatrix}, \tag{1.22}$$

with V_{iT} and I_{iT}, $i = 1, 2$, represent the voltages at and the currents flowing through ports 1 and 2 of the overall structure. Finally, it is worth mentioning that, unlike S-parameters and T-parameters, ABCD-parameters are not defined as a function of a reference impedance.

1.5 S-parameters Measurements

Most of the common tasks in microwave engineering involve the use of electrical measurements at high frequencies. A network analyzer is an instrument that allows to measure S-parameters with great precision. By setting all incident waves to zero in eqn (1.5), except for the incident wave at port j (a_j), one can deduce the simple relationship between a given S-parameters matrix element and a particular ratio of (measurable) scattered to incident waves as

$$S_{ij} = \left. \frac{b_i}{a_j} \right|_{a_k=0, k\neq j}. \tag{1.23}$$

While eqn (1.23) may be used to describe the measurements performed by a multi-port VNA, for simplicity, let us analyze the case of a two-port

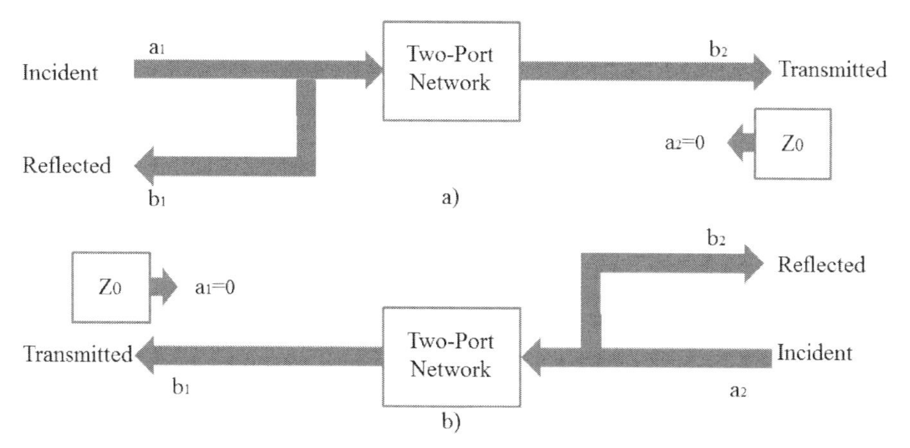

Figure 1.5 S-parameters measurement in (a) forward direction and (b) reverse direction.

network. For a two-port network, the measurement of S-parameters is carried out in two steps: (a) forward direction, in which the incident signal is set at port 1, and (b) reverse direction, in which the signal source is set at port 2, as depicted in Figure 1.5.

In Figure 1.5, the stimulus is an incident wave at port 1. The fact that a_2 is not present ($a_2 = 0$) is interpreted to mean that the b_2 wave scattered and traveling away from port 2 is not reflected back into the device at port 2. Under this condition, the device is said to be perfectly matched at port 2. Two of the four complex S-parameters, namely S_{11} and S_{21}, can be identified from eqn (1.23) as

$$S_{11} = \frac{b_1}{a_1}\bigg|_{a_2=0}, \tag{1.24}$$

$$S_{21} = \frac{b_2}{a_1}\bigg|_{a_2=0}. \tag{1.25}$$

In reverse direction, the stimulus is an incident wave at port 2. In this case, the fact that a_1 is not present ($a_1 = 0$) is interpreted to mean that the b_1 wave scattered and traveling away from port 1 is not reflected back into the device at port 1. Under this condition, the device is said to be perfectly matched at port 1. Two of the four S-parameters, S_{12} and S_{22}, can be identified from eqn (1.23) as

$$S_{22} = \frac{b_2}{a_2}\bigg|_{a_1=0}, \tag{1.26}$$

$$S_{12} = \frac{b_1}{a_2}\bigg|_{a_1=0}. \tag{1.27}$$

In practice, the ratio on eqn (1.23) may vary with the magnitude of the incident wave. Nevertheless, the identification of this ratio with the S-parameters of a device is valid only if the incident wave allows the DUT to operate in its linear region. For nonlinear components, such as transistors or amplifiers, the scattered waves eventually increase nonlinearly with the incident wave as the latter becomes larger in magnitude. A better and more complete definition of S-parameters for a nonlinear component is given by [79]

$$S_{ij} = \lim_{a_j \to 0} \frac{b_i}{a_j} \bigg|_{a_k=0, k \neq j}. \tag{1.28}$$

That is, for a general active device, S-parameters are defined as ratios of output responses to input stimuli in the limit of small input signals. This emphasizes that S-parameters apply to nonlinear components only in the small-signal limit [79].

1.6 Large-signal Measurement Systems

Active devices used for power amplification, frequency multiplication, and mixing are typically operated near to their compression point, in their nonlinear region. Therefore, the measurement of devices in nonlinear operation requires large-signal measurement systems [37, 77]. So far, the most commonly used measurement systems to accomplish this purpose are the nonlinear vector network analyzer (NVNA), the large-signal network analyzer, along with the family of load-pull systems. While the former two types of systems are, or have existed as, commercially available standalone solutions [85], the latter exist in many different solutions, in which passive tuners, active load-pull systems, along with hybrid approaches are usually implemented [33]. In addition, more recent advances in this field are the integration of load-pull systems with nonlinear measurement systems as standalone instruments [50].

This book is intended to present the calibration of small-signal measurements using different calibration techniques, to explain how these systems are used to describe the electrical behavior of microwave devices, and to draw the path from small-signal measurement systems to large-signal measurement systems.

2

Error Models for Vector Network Analyzers

A network analyzer is an instrument that internally encompasses an RF source, a switch to route the signal from the source to either port 1 or port 2 of a DUT, a pair of directional couplers to separate the incident and scattered waves at the DUT ports, and the port receivers (detectors) for measuring those waves, as depicted in Figure 2.1.

Imperfections of the VNA's elements (e.g., ports and source mismatch, low directivity of directional couplers, and losses in transmission paths) along with effects introduced by other elements of the measurement setup (e.g., cables, adapters, connectors, bias tees, and test fixtures) introduce errors in the measurement of the incident and scattered waves at the DUT ports. In order to determine the contribution of these errors and to remove their effects from the DUT's measurements, a calibration procedure has to be carried out.

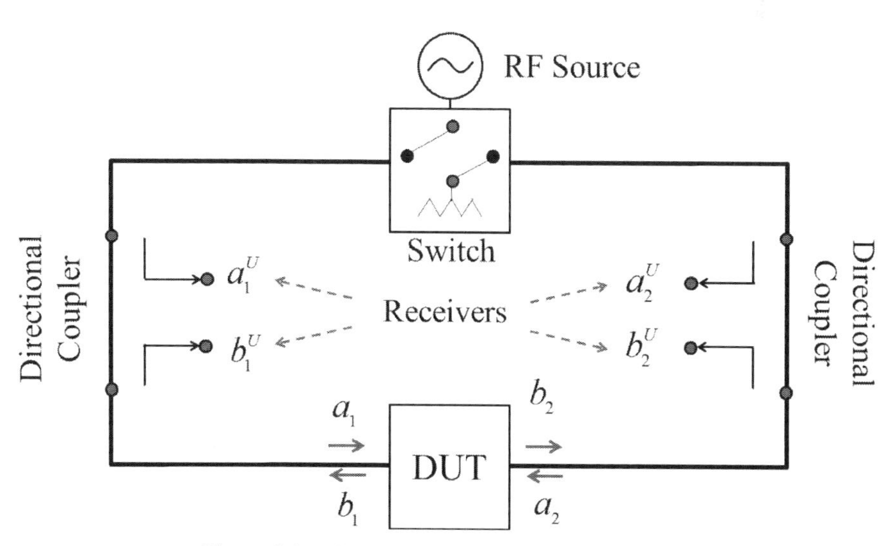

Figure 2.1 Simplified diagram of a two-port VNA.

In a calibration procedure, there are two necessary elements: a calibration technique and an error model. This section is devoted to describe the most commonly used error models in VNA calibration.

The goal of every VNA calibration procedure is to provide the user with an electrical connection having a reference plane set at the DUT input and output ports, and a reference impedance set to the measurement system's impedance, Z_0. Every measurement setup has imperfections that corrupt the measurement; thus, mathematically placing an error model between the VNA and the DUT is a method to represent these imperfections [29].

The calibration serves to identify the error terms between the VNA system and the DUT by using information provided by the measurement of a set of calibration structures connected at the desired calibration plane.

2.1 Three-Terms Error Model

The errors associated with a single port of the VNA can be represented by a two port S-parameter block that is usually referred to as an error box. Because S-parameters are ratio measurements, the four error terms of this portion of the VNA error flow graph can be normalized, thereby reducing the one-port error model to three terms. These three terms are often named as directivity, source match, and reflection frequency tracking.

Directivity is a parameter that pertains to the measurement couplers. The directivity is the difference in the power output at a coupled port, when power is transmitted in a desired direction to the power output at the same coupled port when the same amount of power is transmitted in the opposite direction. Ideally, the VNA's directional couplers in the analyzer would only sample the power of the intended measurement signal. The directivity error term is associated with unwanted leakage in the couplers.

The port match or source match is defined as the vector sum of the signals present at the output of the network analyzer due to its inability to maintain a constant power at the input of the test device. In the forward model, the port 1 match is the source match; conversely, in the reverse model, the port 2 match is the source match. Uncertainty is caused when the impedance of the source does not match the impedance of the port that connects the input of the DUT and the impedance of the analyzer's port does not match the impedance of the DUT input connections. The purpose of the source match term is to reduce this uncertainty.

The reflection frequency tracking is the vector sum of measurement variations in magnitude and phase of the frequency response of the reflected

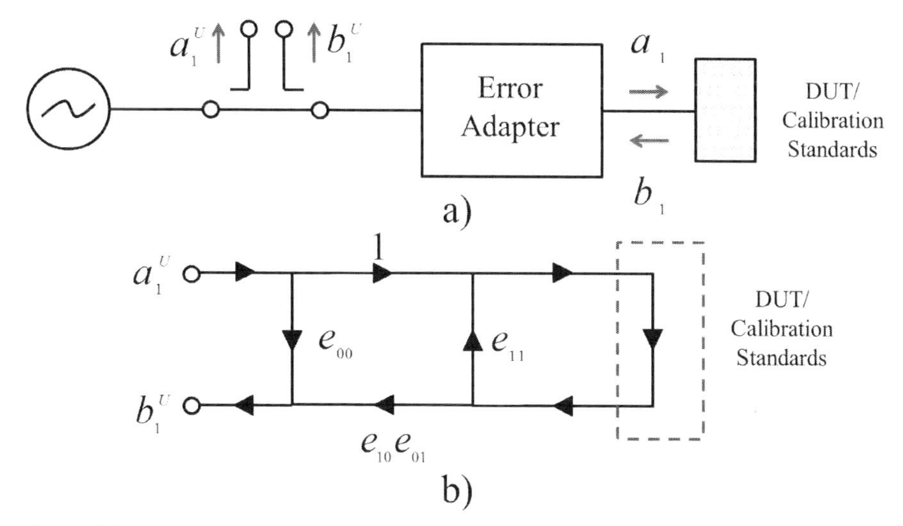

Figure 2.2 (a) VNA configuration for one-port measurements and (b) signal flow diagram of its error model.

signal. Basically, these are the errors, due to the fact that the RF source, the incident sampler, and the reflected sampler are not physically in the same place and are not at the input of the DUT. Therefore, there are losses and phase deviations due to cables and signal flow though the VNA that the reflection tracking term corrects for.

The three-term error model is a model suited for one-port calibration. Figure 2.2(a) shows the VNA configuration and signal flow diagram representation of the three-term error model. In this model, the VNA is represented by a perfect reflectometer with an error adapter between the VNA's receivers and the DUT. The error adapter links the waves $\left(a_1^U, b_1^U\right)$ and reflection coefficient $\left(\Gamma_1^U = b_1^U/a_1^U\right)$ at the plane of the VNA receivers with the waves (a_1, b_1) and reflection coefficient $(\Gamma_1 = b_1/a_1)$ at the DUT plane.

The error adapter contains three error terms, which are referred to as directivity (e_{00}), source match (e_{11}), and reflection response $(e_{10}e_{01})$ errors. From the diagram shown in Figure 2.2(b), the following expression relating Γ_1^U with Γ_1 may be derived:

$$\Gamma_1^U = \frac{b_1}{a_1} = \frac{e_{00} - \Delta_e \Gamma_1}{1 - e_{11}\Gamma_1}. \tag{2.1}$$

After some algebra, the following expression may be obtained:

$$e_{00} - \Delta_e \Gamma_1 + e_{11} \Gamma_1 \Gamma_1^U = \Gamma_1^U. \qquad (2.2)$$

The objective of a one-port calibration is to determine the error terms by using the measurement of (at least) three loads of known reflection coefficient. The most common one-port calibration procedure is short-open-load (SOL), which uses a high-impedance load (open), a low-impedance load (short), and a load of impedance close to the measurement system's impedance (load or match).

2.2 Twelve-Terms Error Model

Unlike one-port measurements, in two-port measurements, there are separate forward and reverse VNA measurements. Figure 2.3 shows the VNA configuration corresponding to the forward and reverse measurements, where independent two-port adapters are used to relate traveling waves at the DUT plane with traveling waves measured at the VNA receivers plane.

Figure 2.4 shows the signal flow diagrams representing the forward and reverse measurements. Each signal flow graph is composed of six terms: directivity (e_{00} and e_{33}), source match (e_{11}^F and e_{22}^R), load match (e_{22}^F and e_{11}^R), transmission response ($e_{10}e_{32}$ and $e_{01}e_{23}$), reflection response ($e_{10}e_{01}$ and $e_{23}e_{32}$), and isolation (e_{30} and e_{03}).

A set of calibration structures of either known or partially known electrical characteristics, typically open circuits, short circuits, matched loads, and the through connection of the VNA ports at the calibration plane have to be measured in order to calculate the 12 error term. Forward and reverse

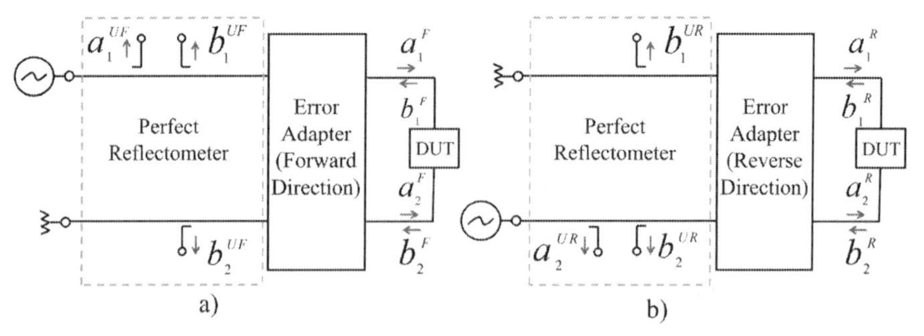

Figure 2.3 VNA configuration diagram representing the (a) forward and (b) reverse measurements.

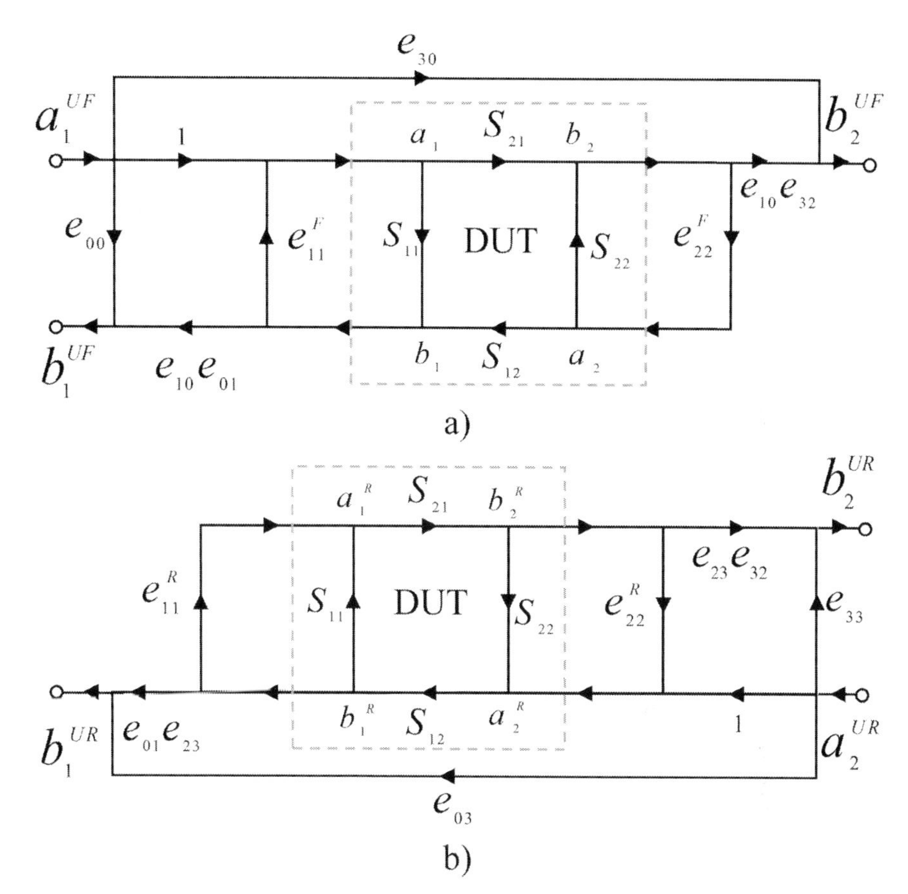

Figure 2.4 Signal flow diagram representing the (a) forward and (b) reverse measurements.

directivity, source match, and reflection response at ports 1 and 2 of the VNA may be determined by using the one-port calibration procedure described in the preceding section at both VNA ports. The connection of the VNA ports through a two-port structure of known electrical behavior is used to determine the forward and reverse transmission response and load match terms; these terms assess the frequency response of the transmission signal path and the impedance of the ports. Finally, the connection of a load of very low reflection coefficient (match) is used to determine the isolation terms, typically in a separate step [92].

In most cases, the isolation terms are smaller than the connector repeatability and can be neglected. Furthermore, note that the forward source match and the reverse load match terms should be identical; similarly, the forward load match and the reverse source match should be identical. These four terms can be reduced to two as long as errors due to mismatches between the source and the terminating load every time the switch, as depicted in Figure 2.3, changes position are removed. These error terms are known as *switching errors*. The following section describes the procedure to correct the switching errors.

2.3 Switching Errors

The accuracy of the vector network analyzer (VNA) is enhanced by calibration. The calibration process of VNA consists of the following steps:

- Removing the switching errors.
- Removing isolation.
- Finally, using the 12-term error model or the 8-term error model along with any of the two-port calibration techniques, the systematic errors are removed and the VNA can be completely calibrated.

In order to accurately determine the S-parameters of any two-port device, a calibration process is required as mentioned in [23]. When the switching errors are not removed, the calibrated S-parameters of the DUT exhibit a noticeable noise, diminishing the reliability of the measurements. The scattering parameters of a two-port device are measured as the ratio of reflected and incident signals when the ports are terminated with Z_0 load, as shown in Figure 2.5. Normally, the reflection S_{11} and the transmission S_{21} parameters are measured when the switch is placed in the forward position as depicted in Figure 2.5. Meanwhile, the reflection S_{22} and transmission S_{12} are measured when the switch is placed in the reverse position as shown in Figure 2.6. Note that the switch changes the direction of incident wave to the unknown two-port for forward and reverse measurements and terminates the unknown two-port device to an impedance Z_0. However, the two-port scattering parameters present errors, due to mismatches between the source and the termination Z_0. The errors produced by mismatches between the source and the termination Z_0 are known as switching errors. The procedure for correcting the switching errors consists of measuring the DUT (device under test) S-parameters, with the system shown in Figures 2.5 and 2.6 for the forward and reverse directions. A detailed procedure for switching errors correction is developed and presented next.

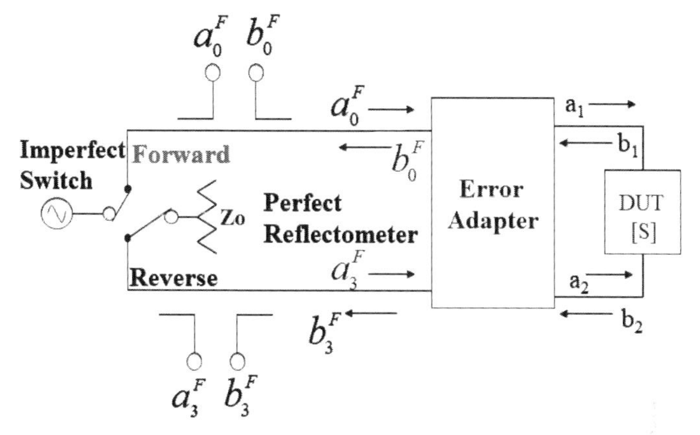

Figure 2.5 VNA diagram under forward direction.

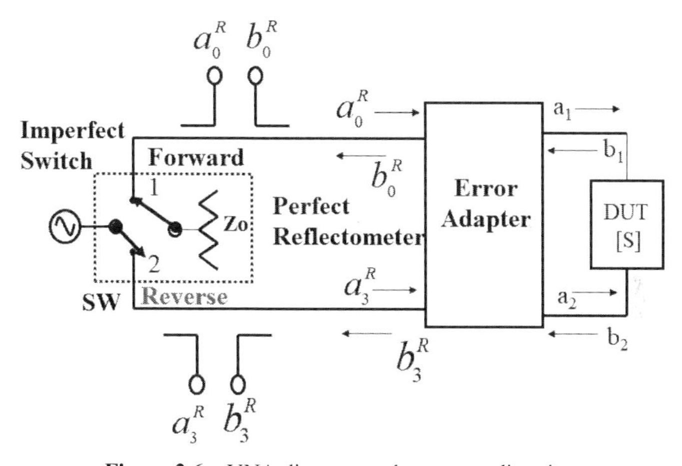

Figure 2.6 VNA diagram under reverse direction.

2.3.1 Forward measurement

Figure 2.5 shows the system configuration used for measuring the DUT (device under test) S-parameters in the system forward configuration. The DUT S-parameters as a function of the measurements $a_0^F, b_0^F, a_3^F, b_3^F$ can be expressed as

$$b_0^F = S_{11M}a_0^F + S_{12M}a_3^F \tag{2.3}$$

$$b_3^F = S_{21M}a_0^F + S_{22M}a_3^F. \tag{2.4}$$

Expressing eqn (2.3) and (2.4) in a matrix form, one has

$$\begin{pmatrix} b_0^F \\ b_3^F \end{pmatrix} = \begin{pmatrix} S_{11M} & S_{12M} \\ S_{21M} & S_{22M} \end{pmatrix} \begin{pmatrix} a_0^F \\ b_3^F \end{pmatrix}. \tag{2.5}$$

2.3.2 Reverse measurement

Figure 2.6 shows the system configuration used for measuring the DUT S-parameters measured in the reverse configuration and expressed as a function of $a_0^R, b_0^R, a_3^R, b_3^R$ by

$$b_0^R = S_{11M}a_0^R + S_{12M}a_3^R \tag{2.6}$$

$$b_3^R = S_{12M}a_0^R + S_{21M}a_3^R. \tag{2.7}$$

Once again, by writing eqn (2.6) and (2.7) in a matrix form, one has

$$\begin{pmatrix} b_0^R \\ b_3^R \end{pmatrix} = \begin{pmatrix} S_{11M} & S_{12M} \\ S_{21M} & S_{22M} \end{pmatrix} \begin{pmatrix} a_0^R \\ b_3^R \end{pmatrix}. \tag{2.8}$$

Combining eqn (2.5) and (2.8), the following matrix equation results:

$$B = [S][A], \tag{2.9}$$

where

$$[B] = \begin{pmatrix} b_0^F & b_0^R \\ b_3^F & b_3^R \end{pmatrix}, \tag{2.10}$$

$$[S] = \begin{pmatrix} S_{11M} & S_{21M} \\ S_{21M} & S_{22M} \end{pmatrix}, \tag{2.11}$$

and

$$[A] = \begin{pmatrix} a_0^F & a_0^R \\ a_3^F & a_3^R \end{pmatrix}. \tag{2.12}$$

Assuming that the matrices in eqn (2.10) and (2.12) are non-singular matrices, the S-matrix is given by

$$S = [B][A]^{-1}. \tag{2.13}$$

Substituting eqn (2.10) and (2.12) into eqn (2.13) and developing, one has

$$\begin{pmatrix} S_{11M} & S_{12M} \\ S_{21M} & S_{22M} \end{pmatrix} = \begin{pmatrix} b_0^F & b_0^R \\ b_3^F & b_3^R \end{pmatrix} \begin{pmatrix} a_0^F & a_0^R \\ a_3^F & a_3^R \end{pmatrix}^{-1}. \tag{2.14}$$

Eqn (2.14) allows determining the DUT S-parameters even if the system has not been terminated with Zo impedance since $a_3^F \neq 0$ in forward configuration and $a_0^R \neq 0$ in reverse configuration. From eqn (2.14), two procedures for removing the switching error may be derived, as presented next.

2.3.3 Procedure I

Procedure I for switching error correction requires the measurements of $a_0^F, b_0^F, a_3^F, b_3^F$ in forward configuration, and the measurements of $a_0^R, b_0^R, a_3^R, b_3^R$ in reverse configuration. Then, developing eqn (2.14) and comparing the left and right terms, the scattering parameters of the DUT are obtained and given by

$$S_{11M} = \frac{b_0^F a_3^R - b_0^R a_3^F}{\Delta} \tag{2.15}$$

$$S_{12M} = \frac{b_0^R a_0^F - b_0^F a_0^R}{\Delta} \tag{2.16}$$

$$S_{21M} = \frac{b_3^F a_3^R - b_3^R a_3^F}{\Delta} \tag{2.17}$$

$$S_{22M} = \frac{b_3^R a_0^F - b_3^F a_0^R}{\Delta} \tag{2.18}$$

$$\Delta = a_0^F a_3^R - a_0^R a_3^F. \tag{2.19}$$

2.3.4 Procedure II

Procedure II is an improvement of procedure I and to do that, eqn (2.19) is expressed as

$$\Delta = a_0^F a_3^R (1 - \frac{a_0^R a_3^F}{a_0^F a_3^R}). \tag{2.20}$$

Next, replacing Δ in eqn (2.15)–(2.18), the DUT S-parameters are calculated by eqn (2.21)–(2.24)

$$S_{11M} = \frac{\frac{b_0^F}{a_0^F} - \frac{b_0^R}{a_3^R}\frac{b_3^F}{a_0^F}\Gamma_2}{d} = \frac{S_{11\text{raw}} - S_{12\text{raw}} S_{21\text{raw}} \Gamma_2}{d} \tag{2.21}$$

$$S_{21M} = \frac{\frac{b_3^F}{a_0^F} - \frac{b_3^R}{a_3^R}\frac{b_3^F}{a_0^F}\Gamma_2}{d} = \frac{S_{21\text{raw}} - S_{22\text{raw}}S_{21\text{raw}}\Gamma_2}{d} \tag{2.22}$$

$$S_{12M} = \frac{\frac{b_0^R}{a_3^R} - \frac{b_0^F}{a_0^F}\frac{b_0^R}{a_3^R}\Gamma_1}{d} = \frac{S_{12\text{raw}} - S_{11\text{raw}}S_{12\text{raw}}\Gamma_1}{d} \tag{2.23}$$

$$S_{22M} = \frac{\frac{b_3^R}{a_3^R} - \frac{b_3^F}{a_0^F}\frac{b_0^R}{a_3^R}\Gamma_1}{d} = \frac{S_{22\text{raw}} - S_{21\text{raw}}S_{12\text{raw}}\Gamma_1}{d} \tag{2.24}$$

where

$$d = 1 - \frac{b_3^F}{a_0^F}\frac{b_0^R}{a_3^R}\frac{a_0^R}{b_0^R}\frac{a_3^F}{b_3^F} = 1 - S_{21\text{raw}}S_{12\text{raw}}\Gamma_1\Gamma_2 \tag{2.25}$$

$$\Gamma_1 = \frac{a_0^R}{b_0^R} \tag{2.26}$$

$$\Gamma_2 = \frac{a_3^F}{b_3^F}. \tag{2.27}$$

In this procedure, the ratios $\Gamma_1 = \frac{a_0^R}{b_0^R}$ and $\Gamma_2 = \frac{a_3^F}{b_3^F}$ are measured first under reverse and forward conditions by connecting a two-port element, as for example, a thru or line between ports 1 and 2 of the vector network analyzer and then the raw S-parameters ($S_{11\text{raw}} = \frac{b_0^F}{a_0^F}$, $S_{21\text{raw}} = \frac{b_3^F}{a_0^F}$, $S_{12\text{raw}}=\frac{b_0^R}{a_3^R}$, $S_{22\text{raw}} = \frac{b_3^R}{a_3^R}$) of the DUT are measured. In summary, to remove the switching errors, using the eqn (2.21) – (2.27), it is not necessary to perform the four measurements in forward $a_0^F, b_0^F, a_3^F, b_3^F$ and the four measurements in reverse configuration $a_0^R, b_0^R, a_3^R, b_3^R$. Furthermore, it is important to mention that the accuracy of the vector network analyzer is enhanced when the ratio of signals is measured. Indeed, procedure II reduce the number of measurements from eight ($a_0^F, b_0^F, a_3^F, b_3^F, a_0^R, b_0^R, a_3^R, b_3^R$) to six ($\frac{b_0^F}{a_0^F} = S_{11\text{raw}}$, $\frac{b_3^F}{a_0^F} = S_{21\text{raw}}$, $\frac{b_0^R}{a_3^R} = S_{12\text{raw}}$, $\frac{b_3^R}{a_3^R} = S_{22\text{raw}}$, Γ_1, Γ_2). Thus, for each DUT S-parameters to be corrected, only six ratios of measurements are needed, but two of them, Γ_1, Γ_2 are independent of the DUT and can be measured once by using a calibration element either a thru or a line. Note that Γ_1 and Γ_2 do not change when the DUT is changed.

2.3.5 Application examples

An application example of how the exposed theory, in this chapter, can be applied to improve the accuracy of the S-parameter data used in the characterization of coaxial transmission lines is given. In Figures 2.7 and in 2.8, the dielectric effective constant versus frequency and the loss constant versus frequency of a coaxial airline are plotted respectively.

It is observed from these plots that when the switching errors have not been corrected, the traces of both the dielectric constant and loss constant are very noisy.

The S-parameters of a two-port DUT are measured by capturing the ratio of the scattered to the incident waves at the DUT ports when one of the ports is terminated with a load of impedance Z_0. In ideal conditions, the instrument's switch routes the signal from the source to one of the DUT ports while terminates the other port with a load of impedance Z_0. However, due to mismatches between the source and the terminating

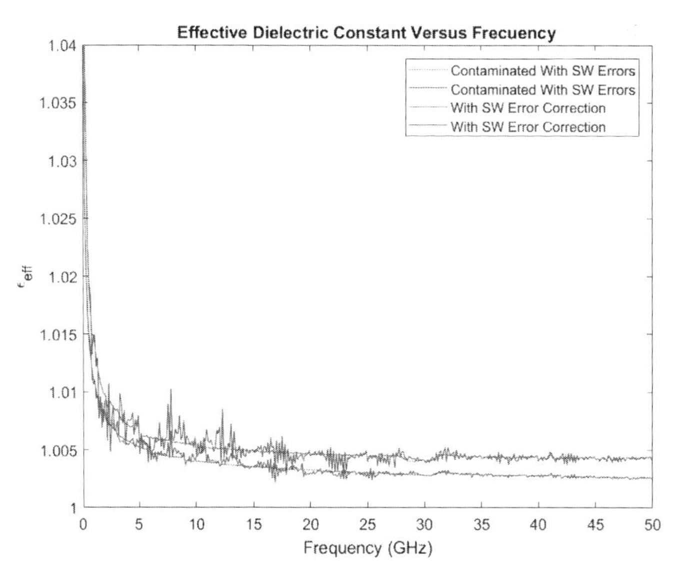

Figure 2.7 Dielectric effective constant versus frequency (Maury model 7943H and 7943G).

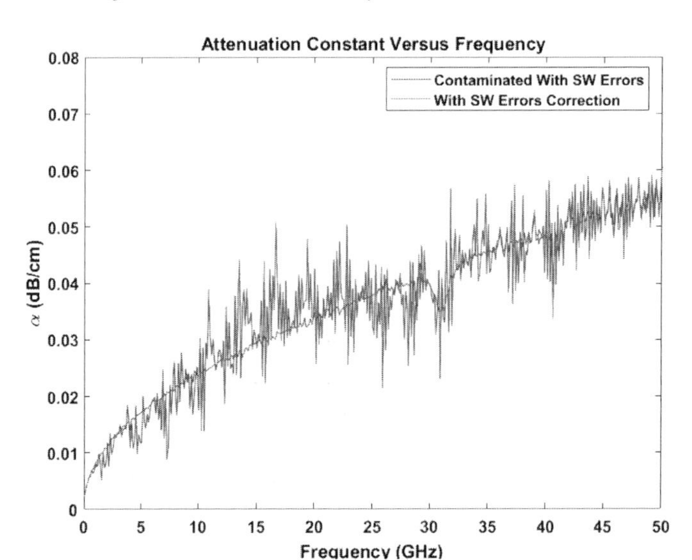

Figure 2.8 Attenuation constant versus frequency (Maury model 7943H).

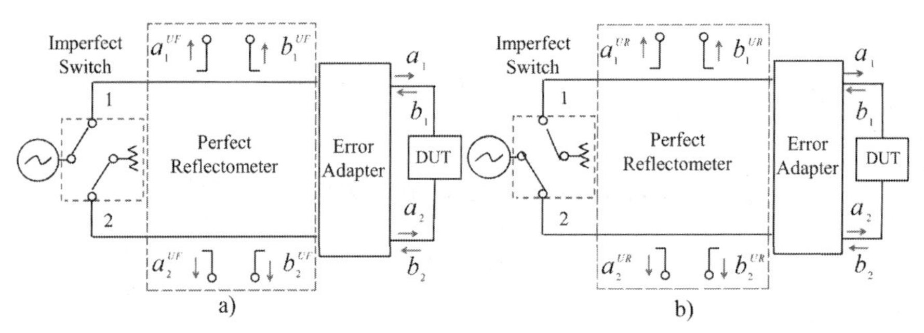

Figure 2.9 VNA configuration for the correction of the switching errors in the (a) forward and (b) reverse directions.

load every time the switch changes position, the measured S-parameters are erroneous.

2.4 Eight-Terms Error Model

Once the switching errors are removed and the isolation errors are either neglected or removed before the calibration procedure, the 12-term error model is reduced to an error model comprising eight terms, four at each port. As shown in Figure 2.10, it allows representing the measurement system as

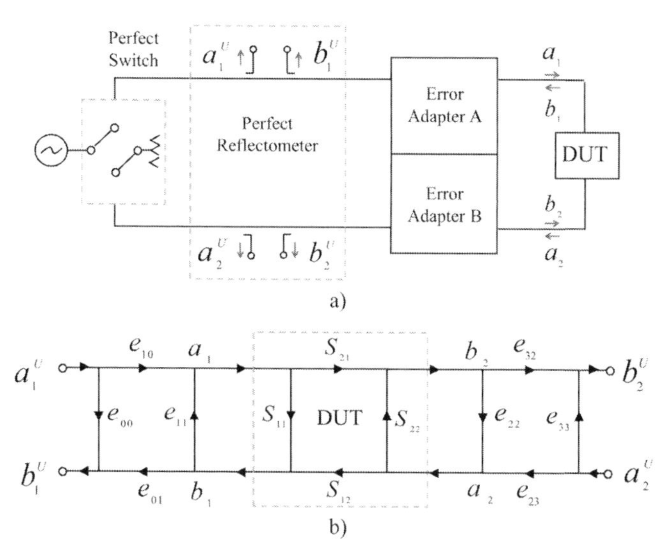

Figure 2.10 (a) VNA configuration diagram and (b) signal flow diagram of the eight-term error model.

the cascade connection of three two-port networks, representing the DUT and error adapters at each VNA port.

The representation of the VNA as the cascade connection of two-port networks allows using mathematical tools for circuit analysis, such as the T-parameters and ABCD-parameters to analyze the measurements performed in a two-port VNA. As a consequence, advanced calibration techniques, such as TRL, TRM, or TRRM, may be implemented to determine the VNA systematic errors. The advantage of using this family of calibration techniques is that the calibration structures used to determine the error terms do not have to be fully known and some of their electrical characteristics are self-calculated in the calibration process. The mathematical procedures of these techniques are fully described in Chapter 4.

3

Microwave De-embedding

3.1 Introduction

Having corrected the switching errors due to non-ideal source and load match along with the isolation between ports, the eight-term error model for modeling the imperfections of the vector network analyzer (VNA) has been used successfully for a long time [21, 22]. Using the 8-term error model [39] along with any of the self-calibration techniques [21, 22], the VNA can be appropriately calibrated. Once the VNA is calibrated, the next step is to measure the scattering parameters of the devicc under test (DUT), whereby a de-embedding process is needed. A classical de-embedding [21, 22] method requires the knowledge of at least seven error terms, which are: a, b, c, α, β, φ, and $\mathrm{r}_{22}\rho_{22}$ [21]. In this chapter, using the eight-term error model and a novel matrix approach reported in [45], the straightforward de-embedding method for coaxial, in-fixture, and on-wafer devices is presented. The advantage of the straightforward de-embedding method over the classical de-embedding method is the use of only three error terms: $a_m/\mathrm{c_m}$, $\mathrm{b_m}$, and a_m rather than the seven error terms used by the classical de-embedding method. This chapter is organized as follows: the de-embedding and the straightforward de-embedding method are presented and different procedures for computing $a_m/\mathrm{c_m}$, $\mathrm{b_m}$, and a_m are presented.

3.2 The General De-embedding Method

3.2.1 The test fixture model

It will be assumed that the device under test, DUT is embedded in uniform transmission lines, as shown in Figure 3.1. By using the wave cascading matrix, WCM, formalism, the resultant $\mathbf{T_D}$ matrix of the three two-port elements connected in a cascade of Figure 3.1 is equal to the product of the

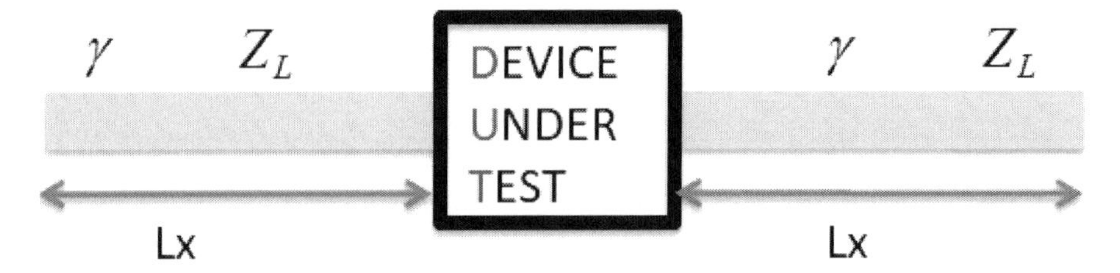

Figure 3.1 Test fixture.

three individual matrices expressed as

$$\mathbf{T_D} = \mathbf{T_L}\,\mathbf{T_{DUT}}\,\mathbf{T_L}. \tag{3.1}$$

In eqn (3.1), $\mathbf{T_L}$ is the WCM of a uniform line where the DUT is embedded and defined as

$$\mathbf{T_L} = \frac{1}{1-\Gamma^2}\frac{1}{e^{-\gamma L_x}}\begin{pmatrix} e^{-2\gamma L_x} - \Gamma^2 & \Gamma(1-e^{-2\gamma L_x}) \\ -\Gamma(1-e^{-2\gamma L_x}) & 1-\Gamma^2 e^{-2\gamma L_x} \end{pmatrix}, \tag{3.2}$$

where Γ is the reflection coefficient given as

$$\Gamma = \frac{Z_L - Z_{\mathrm{ref}}}{Z_L + Z_{\mathrm{ref}}}, \tag{3.3}$$

where Z_L is the line characteristic impedance and Z_{ref} is the system reference impedance.

In eqn (3.2), γ is the propagation constant and L_x is the physical length of the line where the DUT is embedded, and $\mathbf{T_{DUT}}$ is the WCM of the DUT defined as

$$\mathbf{T_{DUT}} = \frac{1}{S_{21}}\begin{pmatrix} -\Delta S & S_{11} \\ -S_{22} & 1 \end{pmatrix}; \qquad \Delta S = S_{11}S_{22} - S_{12}S_{21}. \tag{3.4}$$

Moreover, eqn (3.2) may be expressed as

$$\mathbf{T_L} = \mathbf{T_\Gamma}\,\mathbf{T_\Lambda}\,\mathbf{T_\Gamma}^{-1}. \tag{3.5}$$

In eqn (3.5), $\mathbf{T_\Gamma}$ and $\mathbf{T_\Lambda}$ are given by

$$\mathbf{T_\Gamma} = \begin{pmatrix} 1 & \Gamma \\ \Gamma & 1 \end{pmatrix}. \tag{3.6}$$

$$\mathbf{T_\Lambda} = \begin{pmatrix} e^{-\gamma L_x} & 0 \\ 0 & e^{\gamma L_x} \end{pmatrix}, \tag{3.7}$$

Substituting eqn (3.5) into eqn (3.1), one has

$$\mathbf{T_D} = \mathbf{T_\Gamma}\, \mathbf{T_\Lambda}\, \widetilde{\mathbf{T}}_{\mathbf{DUT}} \mathbf{T_\Lambda}\, \mathbf{T_\Gamma}^{-1}, \tag{3.8}$$

where

$$\widetilde{\mathbf{T}}_{\mathbf{DUT}} = \mathbf{T}_\Gamma^{-1} \mathbf{T_{DUT}} \mathbf{T_\Gamma}. \tag{3.9}$$

$\widetilde{\mathbf{T}}_{\mathbf{DUT}}$ is defined as

$$\widetilde{\mathbf{T}}_{\mathbf{DUT}} = \frac{1}{\widetilde{S}_{21}} \begin{pmatrix} -\Delta\widetilde{S} & \widetilde{S}_{11} \\ -\widetilde{S}_{22} & 1 \end{pmatrix} \qquad \Delta\widetilde{S} = \widetilde{S}_{11}\widetilde{S}_{22} - \widetilde{S}_{12}\widetilde{S}_{21}. \tag{3.10}$$

In eqn (3.9), $\widetilde{\mathbf{T}}_{\mathrm{DUT}}$ is the DUT matrix modified by the line impedance. Note that when the line is non-reflecting, the product $\widetilde{\mathbf{T}}_{\mathbf{DUT}} = \mathbf{T}_\Gamma^{-1} \mathbf{T_{DUT}} \mathbf{T_\Gamma}$ simplifies in $\mathbf{T_{DUT}}$.

3.3 The Vector Network Analyzer Model

For modeling the imperfect measurement system that includes the cables, the connectors, the test fixture, and the VNA, the eight-term model is commonly used and is shown in Figure 3.2. The error box is denoted as A and B in Figure 3.2 and includes the fixture errors along with the error terms of the VNA. The wave cascade matrix WCM is used for modeling the error boxes A and B. The WCM of the error boxes A and B is defined as $\mathbf{T_A}$ and $\mathbf{T_B}$, respectively, and are defined as [21]

$$\mathbf{T_A} = r_{22} \begin{pmatrix} a & b \\ c & 1 \end{pmatrix}, \tag{3.11}$$

and

$$\mathbf{T_B} = \rho_{22} \begin{pmatrix} \alpha & \beta \\ \varphi & 1 \end{pmatrix}. \tag{3.12}$$

3.4 The Classical De-embedding Method

The classical de-embedding method requires, for their implementation, that $\mathbf{T_A}$ and $\mathbf{T_B}$ matrix be known. These matrices may be determined using any

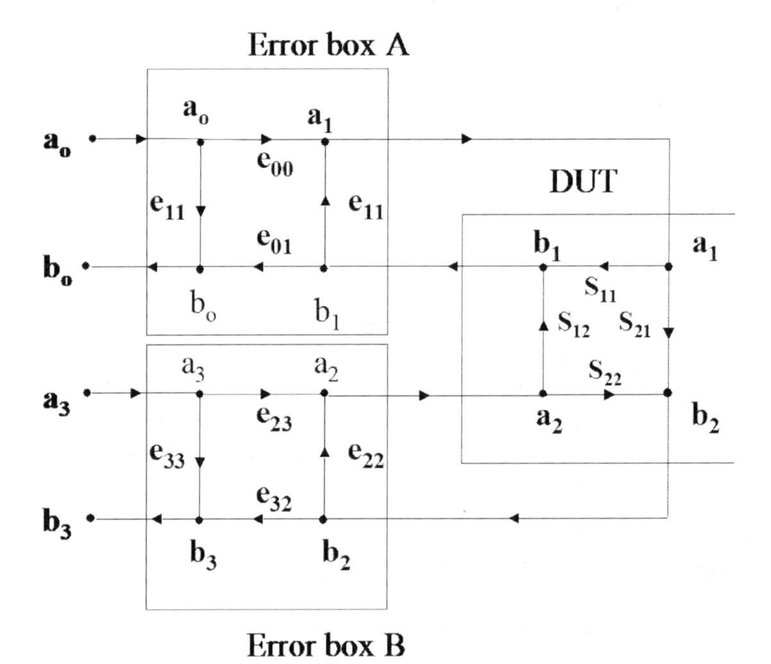

Figure 3.2 Eight-term error model [29].

two-port calibration techniques. The equivalent WCM matrix M_{DUT} resulting when the DUT is connected between ports 1 and 2 of the measurement system is given as

$$\mathbf{M_{DUT}} = \mathbf{T_A}\,\mathbf{T_D}\,\mathbf{T_B}, \tag{3.13}$$

where $\mathbf{T_D}$ is given by eqn (3.8) and expressed in the function of $\mathbf{M_{DUT}}$ as

$$\mathbf{T_A}^{-1}\mathbf{M_{DUT}}\mathbf{T_B^{-1}} = \mathbf{T_\Gamma}\,\mathbf{T_\Lambda}\,\widetilde{\mathbf{T}}_{\mathbf{DUT}}\,\mathbf{T_\Lambda}\,\mathbf{T_\Gamma}^{-1}. \tag{3.14}$$

Next, developing eqn (3.14) and solving for $\widetilde{\mathbf{T}}_{\mathbf{DUT}}$, one has

$$\widetilde{\mathbf{T}}_{\mathbf{DUT}} = \mathbf{T_\Lambda^{-1}}\,\mathbf{T_x^{-1}}\,\mathbf{M_{DUT}}\,\mathbf{T_y^{-1}}\mathbf{T_\Lambda^{-1}}. \tag{3.15}$$

In eqn (3.15), $\mathbf{T_x}$ and $\mathbf{T_y}$ are defined as: $\mathbf{T_x} = \mathbf{T_A}\,\mathbf{T_\Gamma}, \mathbf{T_y} = \mathbf{T_\Gamma^{-1}}\,\mathbf{T_B}$ and expressed by

$$\mathbf{T_x} = \mathbf{T_A}\,\mathbf{T_\Gamma} = r_{22}\,(c\,\Gamma + 1)\begin{pmatrix} \frac{a+b\,\Gamma}{1+c\,\Gamma} & \frac{a\,\Gamma+b}{1+c\,\Gamma} \\ \frac{c+\Gamma}{1+c\,\Gamma} & 1 \end{pmatrix} = r_m \begin{pmatrix} a_m & b_m \\ c_m & 1 \end{pmatrix}, \tag{3.16}$$

$$\mathbf{T_y} = \mathbf{T_\Gamma^{-1}}\,\mathbf{T_B} = \rho_{22}\,\frac{1+\beta\,\Gamma}{1-\Gamma^2}\begin{pmatrix} \frac{\alpha+\varphi\,\Gamma}{1-\beta\,\Gamma} & \frac{\beta-\Gamma}{1-\beta\,\Gamma} \\ \frac{\varphi+\alpha\,\Gamma}{1-\beta\,\Gamma} & 1 \end{pmatrix} = \rho_m \begin{pmatrix} \alpha_m & \beta_m \\ \varphi_m & 1 \end{pmatrix}. \tag{3.17}$$

$\mathbf{T_\Gamma}$ is the impedance eigenvector matrix and $\mathbf{T_x}$ and $\mathbf{T_y}$ are the new error matrix modified by the line impedance. In that sense, the new calibration error terms a_m, b_m, c_m, r_m, α_m, β_m, φ_m, and ρ_m depend on the line characteristic impedance Z_L, the reference impedance of the system Z_{ref} and the error terms of the vector network analyzer. The Eqn (3.15) will be referred to as the classical de-embedding equation. It is interesting to mention that the DUT S-parameters calculated using eqn (3.15) require the evaluation of the wave propagation constant, γ and the seven error terms (a_m, b_m, c_m, r_m, α_m, β_m, φ_m, and ρ_m), which may be determined by using two-port calibration techniques.

3.5 The Straightforward De-embedding Method [45], [70]

Figure 3.3 shows the structures used in the implementation of the straightforward de-embedding method. It utilizes the S-parameters measurement of a uniform line, referred to as the reference line, and the S-parameters measurement of the DUT both measured with an uncalibrated vector network analyzer. An appropriate algebraic matrix manipulation of these measurements allows us to determine the DUT S-parameters that depend only on the error terms of the matrix $\mathbf{T_x}$. The equivalent WCM matrix $\mathbf{M_{DUT}}$ resulting from the measurement of the DUT is given by eqn (3.13) and the matrix $\mathbf{M_1}$ resulting from the connection of the reference line L_1 may be written as

$$\mathbf{M_1} = \mathbf{T_A}\,\mathbf{T_{L1}}\,\mathbf{T_B}, \tag{3.18}$$

where

$$\mathbf{T_{L1}} = \mathbf{T_\Gamma}\,\mathbf{T_{\lambda_1}}\,\mathbf{T_\Gamma}^{-1}, \tag{3.19}$$

$$\mathbf{T_{\lambda_1}} = \begin{pmatrix} e^{-\gamma L_{\text{ref}}} & 0 \\ 0 & e^{\gamma L_{\text{ref}}} \end{pmatrix}, \tag{3.20}$$

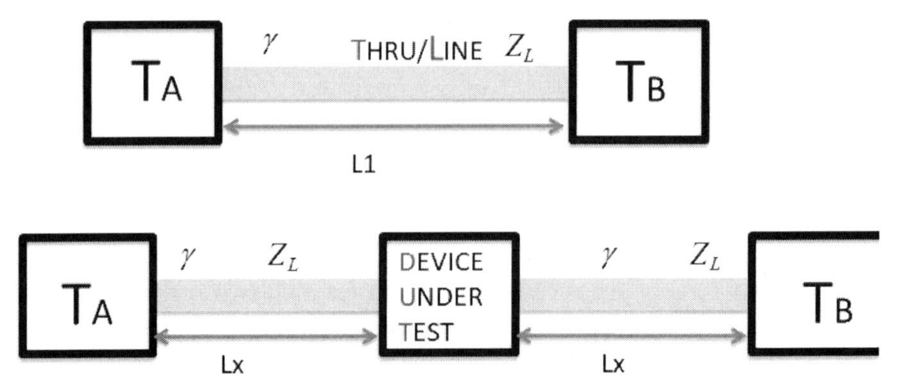

Figure 3.3 Structures used in the straightforward de-embedding method.

As mentioned earlier, $\mathbf{T_{L1}}$ is the WCM of a uniform line of length equal to L_{ref}, $\mathbf{T_D}$ is the WCM of the DUT embedded in lines of length L_x, and $\mathbf{M_{DUT}}$ is the WCM of the DUT given by eqn (3.13). Utilizing eqn (3.13) and eqn (3.18) for calculating the product $\mathbf{M_{DUT}M_1^{-1}}$, eqn (3.21) results

$$\mathbf{M_{DUT}\,M_1^{-1}} = \mathbf{T_A\,T_D\,T_{L1}^{-1}\,T_A^{-1}}. \tag{3.21}$$

Next, solving for $\mathbf{T_D\,T_{L1}^{-1}}$, it is possible to express eqn (3.22) as

$$\mathbf{T_D\,T_{L1}^{-1}} = \mathbf{T_A^{-1}\,M_{DUT}\,M_1^{-1}\,T_A},, \tag{3.22}$$

where the product $\mathbf{M_{DUT}\;M_1^{-1}}$ is defined by eqn (3.23) and expressed as

$$\mathbf{M_{DUT}\,M_1^{-1}} = \begin{pmatrix} t_{11} & t_{12} \\ t_{21} & t_{22} \end{pmatrix}. \tag{3.23}$$

Now, using $\mathbf{T_D}$ given by eqn (3.8) and $\mathbf{T_{L1}}$ given by eqn (3.19), eqn (3.22) becomes

$$\mathbf{T_\Lambda\,\widetilde{T}_{DUT}T_\Lambda\,T_{\lambda_1}} = \mathbf{T_x^{-1}\,M_{DUT}\,M_1^{-1}\,T_x}. \tag{3.24}$$

Then, developing the left-hand term of eqn (3.24), the term $\mathbf{T_\Lambda\,\widetilde{T}_{DUT}\,T_\Lambda\,T_{\lambda_1}^{-1}}$ may be expressed by

$$\mathbf{T_\Lambda\,\widetilde{T}_{DUT}\,T_\Lambda\,T_{\lambda 1}^{-1}} = \frac{1}{\widetilde{S}_{21}} \begin{pmatrix} -\Delta\widetilde{S}\; e^{\gamma(L_{\text{ref}}-2L_x)} & \widetilde{S}_{11}e^{-\gamma\,L_{\text{ref}}} \\ -\widetilde{S}_{22}e^{\gamma\,L_{\text{ref}}} & e^{-\gamma(L_{\text{ref}}-2L_x)} \end{pmatrix}, \tag{3.25}$$

and developing the right-hand term of eqn (3.24), the term $\mathbf{T_x^{-1}}\,\mathbf{M_{DUT}}\,\mathbf{M_1^{-1}}$ $\mathbf{T_x}$ may be given as

$$\mathbf{T_x^{-1}}\,\mathbf{M_{DUT}}\,\mathbf{M_1^{-1}}\,\mathbf{T_x} = \frac{1}{a_m - b_m c_m}\begin{pmatrix} A_{11} & A_{12} \\ A_{21} & A_{22} \end{pmatrix}, \tag{3.26}$$

where

$$A_{11} = a_m\left(t_{11} + \frac{c_m}{a_m}\,t_{12} - b_m\,t_{21} - b_m\,\frac{c_m}{a_m}\,t_{22}\right), \tag{3.27}$$

$$A_{12} = b_m\,t_{11} + t_{12} - b_m^2\,t_{21} - b_m\,t_{22}, \tag{3.28}$$

$$A_{21} = a_m^2\left(-\frac{c_m}{a_m}\,t11 - \left(\frac{c_m}{a_m}\right)^2 t_{12} + t_{21} + \frac{c_m}{a_m}\,t_{22}\right), \tag{3.29}$$

$$A_{22} = a_m\left(-b_m\,\frac{c_m}{a_m}\,t_{11} - \frac{c_m}{a_m}\,t_{12} + b_m\,t_{21} + t_{22}\right). \tag{3.30}$$

Comparing term by term the left-hand side with the right-hand side of eqn (3.24) expressed by eqn (3.25) and eqn (3.26), respectively, the DUT S-parameters $\widetilde{S}_{11}$, $\widetilde{S}_{22}$, and $\widetilde{S}_{21}$ may be expressed analytically. From eqn (3.24), it should be noticed that, $\mathbf{T_A}\widetilde{\mathbf{T}}_{\mathbf{DUT}}\,\mathbf{T_A}\mathbf{T}_{\lambda 1}^{-1}$ is similar to matrix product $\mathbf{M_{DUT}}\mathbf{M_1^{-1}}$, and as a result, they have the same determinant, $\det\left\{\mathbf{T_A}\widetilde{\mathbf{T}}_{\mathbf{DUT}}\,\mathbf{T_A}\mathbf{T}_{\lambda_1}^{-1}\right\} = \det\left\{\mathbf{M_{DUT}}\mathbf{M_1^{-1}}\right\}$, and allow to express analytically $\widetilde{S}_{12}$ by eqn (3.31) as

$$\widetilde{S}_{12} = \widetilde{S}_{21}\Delta t; \quad \Delta t = t_{11}t_{22} - t_{12}t_{22}. \tag{3.31}$$

Finally, the DUT S-parameters may be calculated using the expressions gathered in Table 3.1.

Equations reported in Table 3.1 will be named the general de-embedding equations because they allow the calculation of the DUT S-parameters embedded in arbitrary uniform lines. Furthermore, depending on whether or not the DUT is embedded in a test fixture, equations reported in Table 3.1 may be simplified. It is very important to comment that the right-hand side of eqn (3.26) depends on the measured $\mathbf{t_{ij}}$ parameters and the a_m, a_m/c_m, and b_m elements (unknowns) of the error box A. By contrast, the right-hand side of eqn (3.25) depends on the DUT S-parameters, the wave propagation constant γ of the lines, the mechanical dimensions of the line where the device is embedded, and the mechanical dimension of the reference line. The expressions reported in

Table 3.1 may be simplified when: (a) $L_{\text{ref}} = L_x = 0$, (b) $L_{\text{ref}} = 2L_x$, (c) $L_{\text{ref}} \neq 0$; $L_x = 0$. Examples of the use of the straightforward de-embedding method, for de-embedding coaxial devices, in-fixture devices, and on-wafer devices will be presented next.

3.5.1 Coaxial devices ($L_{\text{ref}} = L_x = 0$)

This case happens when the length of the reference line (L_{ref}) and the line length where the device is embedded (L_X) are both zero, that is, $L_{\text{ref}} = L_x = 0$. This special case corresponds to coaxial devices, and eqn (3.25) becomes

$$\mathbf{T_A}\,\widetilde{\mathbf{T}}_{\mathbf{DUT}}\,\mathbf{T_A}\,\mathbf{T_{A_1}}^{-1} = \frac{1}{\widetilde{S}_{21}}\begin{pmatrix} -\Delta\widetilde{S} & \widetilde{S}_{11} \\ -\widetilde{S}_{22} & 1 \end{pmatrix}, \tag{3.32}$$

and the equations reported in Table 3.1 becomes the expression reported in Table 3.2.

It is important to mention that while the transmission, forward and reverse parameters $\widetilde{S}_{12}$ and $\widetilde{S}_{21}$ are dependent on $\frac{a_m}{c_m}$, and b_m terms, the reflection

Table 3.1 General expressions for calculating the S-parameters of the DUT after the straightforward de-embedding method.

$\widetilde{S}_{11} = \frac{-a_m}{c_m}\frac{b_m^2 t_{21}+b_m(t_{22}-t_{12})}{a_{mx}(\frac{a_m}{c_m}t_{22}-b_m(t_{11}-\frac{a_m}{c_m}t_{11})-t_{12})} e^{-\gamma(L_{\text{ref}}-2L_x)}$	$\widetilde{S}_{12} = \Delta t\left[\frac{\frac{a_m}{c_m}-b_m}{(\frac{a_m}{c_m}t_{22}-b_m(t_{11}-\frac{a_m}{c_m}t_{11})-t_{12})} e^{-\gamma(L_{\text{ref}}-2L_x)}\right]$
$\widetilde{S}_{21} = \frac{\frac{a_m}{c_m}-b}{(\frac{a_m}{c_m}t_{22}-b_m(t_{11}-\frac{a_m}{c_m}t_{11})-t_{12})} e^{-\gamma(L_{\text{ref}}-2L_x)}$	$\widetilde{S}_{22} = \frac{(\frac{a_m}{c_m})^2 t_{21}+b_m(t_{22}-t_{12})}{(\frac{a_m}{c}t_{22}-b_m(t_{11}-\frac{a_m}{c_m}t_{11})-t_{12})} a_{mx} e^{-\gamma(L_{\text{ref}}-2L_x)}$
$a_{mx} = a_m\, e^{\gamma(L_{\text{ref}}-L_x)}$	$\Delta t = t_{11}t_{22} - t_{12}t_{22}$

Table 3.2 Expressions for calculating the DUT S-parameters after the straightforward de-embedding method when $L_{\text{ref}} = 0$ and $L_x = 0$ [97].

$\widetilde{S}_{11} = \frac{-a_m}{c_m}\frac{b_m^2 t_{21}+b_m(t_{22}-t_{12})}{a_{mx}(\frac{a_m}{c_m}t_{22}-b_m(t_{11}-\frac{a_m}{c_m}t_{11})-t_{12})}$	$\widetilde{S}_{12} = \Delta t\left[\frac{\frac{a_m}{c_m}-b_m}{(\frac{a_m}{c_m}t_{22}-b_m(t_{11}-\frac{a_m}{c_m}t_{11})-t_{12})}\right]$
$\widetilde{S}_{21} = \frac{\frac{a_m}{c_m}-b}{(\frac{a_m}{c_m}t_{22}-b_m(t_{11}-\frac{a_m}{c_m}t_{11})-t_{12})}$	$\widetilde{S}_{22} = \frac{(\frac{a_m}{c_m})^2 t_{21}+b_m(t_{22}-t_{12})}{(\frac{a_m}{c}t_{22}-b_m(t_{11}-\frac{a_m}{c_m}t_{11})-t_{12})} a_{mx}$
$a_{mx} = a_m$	$\Delta t = t_{11}t_{22} - t_{12}t_{22}$

parameters $\widetilde{S}_{11}$, $\widetilde{S}_{22}$ are both dependent on $\frac{a_m}{c_m}$, b_m, and a_{mx} terms. In summary, the equations reported in Table 3.2 indicate that four S-parameters of an unknown two-port device may be determined from the knowledge of the $\frac{a_m}{c_m}$, b_m, and a_{mx} terms.

3.5.2 Devices mounted in test fixtures ($L_{\text{ref}} = 2L_x$)

For devices mounted in test fixtures, the length of the reference line is two times the length where the device is embedded, that is, $L_{\text{ref}} = 2L_x$. This case corresponds to the DUT mounted in a test fixture, and the equations reported in eqn (3.25) and in Table 3.2 can be simplified. In this special case, eqn (3.25) becomes

$$\mathbf{T_A}\,\widetilde{\mathbf{T}}_{\mathbf{DUT}}\,\mathbf{T_A}\,\mathbf{T}_{\mathbf{A_1}}^{-1} = \frac{1}{\widetilde{S}_{21}}\begin{pmatrix} -\Delta\widetilde{S} & \widetilde{S}_{11}e^{-\gamma L_{\text{ref}}} \\ -\widetilde{S}_{22}e^{\gamma L_{\text{ref}}} & 1 \end{pmatrix}, \tag{3.33}$$

and the DUT S-parameters are calculated by using the expressions reported in Table 3.3.

Table 3.3 Expressions for calculating the S-parameters of the DUT after the straightforward de-embedding method when $L_x = \frac{L_{\text{ref}}}{2}$ [97].

$\widetilde{S}_{11} = \frac{-a_m}{c_m}\frac{b_m^2 t_{21}+b_m(t_{22}-t_{12})}{a_{mx}(\frac{a_m}{c_m}t_{22}-b_m(t_{11}-\frac{a_m}{c_m}t_{11})-t_{12})}$	$\widetilde{S}_{12} = \Delta t\left[\frac{\frac{a_m}{c_m}-b_m}{(\frac{a_m}{c_m}t_{22}-b_m(t_{11}-\frac{a_m}{c_m}t_{11})-t_{12})}\right]$
$\widetilde{S}_{21} = \frac{\frac{a_m}{c_m}-b}{(\frac{a_m}{c_m}t_{22}-b_m(t_{11}-\frac{a_m}{c_m}t_{11})-t_{12})}$	$\widetilde{S}_{22} = \frac{(\frac{a_m}{c_m})^2 t_{21}+b_m(t_{22}-t_{12})}{(\frac{a_m}{c}t_{22}-b_m(t_{11}-\frac{a_m}{c_m}t_{11})-t_{12})}a_{mx}$
$a_{mx} = a_m\,e^{\gamma L_x}$	$\Delta t = t_{11}t_{22} - t_{12}t_{22}$

It is worth noting that when $L_{\text{ref}} = 2L_x$ (device embedded in uniform lines), the calculations of $\widetilde{S}_{11}$ and $\widetilde{S}_{22}$ parameters using equations reported in Table 3.3 require the knowledge of $\frac{a_m}{c_m}$, b_m, and a_{mx} terms. Concerning the $\widetilde{S}_{12}$ and $\widetilde{S}_{21}$ parameters, it is observed that only $\frac{a_m}{c_m}$, b_m, and terms are required for its calculation.

Table 3.4 Expressions for calculating the S-parameters of the DUT after the straightforward de-embedding method when $L_x = 0$; $L_{\mathrm{ref}} \neq 0$, [97].

$\widetilde{S}_{11} = \frac{-a_m}{c_m} \frac{b_m^2 t_{21} + b_m (t_{22} - t_{12})}{a_{mx}(\frac{a_m}{c_m} t_{22} - b_m (t_{11} - \frac{a_m}{c_m} t_{11}) - t_{12})} e^{-\gamma(L_{\mathrm{ref}})}$	$\widetilde{S}_{12} = \Delta t \left[\frac{\frac{a_m}{c_m} - b_m}{(\frac{a_m}{c_m} t_{22} - b_m (t_{11} - \frac{a_m}{c_m} t_{11}) - t_{12})} e^{-\gamma(L_{ref})} \right]$
$\widetilde{S}_{21} = \frac{\frac{a_m}{c_m} - b}{(\frac{a_m}{c_m} t_{22} - b_m (t_{11} - \frac{a_m}{c_m} t_{11}) - t_{12})} e^{-\gamma(L_{ref})}$	$\widetilde{S}_{22} = \frac{(\frac{a_m}{c_m})^2 t_{21} + b_m (t_{22} - t_{12})}{(\frac{a_m}{c} t_{22} - b_m (t_{11} - \frac{a_m}{c_m} t_{11}) - t_{12})} a_{mx} e^{-\gamma(L_{ref})}$
$a_{mx} = a_m \, e^{\gamma(L_{ref})}$	$\Delta t = t_{11} t_{22} - t_{12} t_{22}$

3.5.3 On-wafer calibration: calibration at the end of the probe's tip ($L_x = 0$; $L_{\mathrm{ref}} \neq 0$)

In applications where the DUT is provided with footprints matching the probe tips, the line where the DUT is embedded is zero, i.e., $L_x = 0$. In this case, the length of the reference line is non-zero. Commercially available calibration standards are provided with a short (typically **1 ps**) reference line. In this case, eqn (3.25) becomes

$$\mathbf{T}_{\Lambda}\, \widetilde{\mathbf{T}}_{\mathbf{DUT}}\, \mathbf{T}_{\Lambda}\, \mathbf{T}_{\Lambda_1}^{-1} = \frac{1}{\widetilde{S}_{21}} \begin{pmatrix} -\Delta \widetilde{S}\, e^{\gamma L_{\mathrm{ref}}} & \widetilde{S}_{11} e^{-\gamma\, L_{\mathrm{ref}}} \\ -\widetilde{S}_{22} e^{\gamma\, L_{\mathrm{ref}}} & e^{-\gamma L_{\mathrm{ref}}} \end{pmatrix}, \tag{3.34}$$

and the DUT S-parameters are calculated by using the expressions reported in Table 3.4, For on-wafer calibration (the reference plane of the calibration is located at the end of the probe tips), the calculation of the DUT S-parameters is done by using the expressions reported in Table 3.4 and requires the knowledge of $\frac{a_m}{c_m}$, b_m, and a_{mx} terms and $e^{\gamma(L_{\mathrm{ref}})}$.

In summary, the calculation of the DUT S-parameters by using the straightforward de-embedding method requires only the knowledge of the $\frac{a_m}{c_m}$, b_m, and a_{mx} terms for coaxial devices, $\frac{a_m}{c_m}$, b_m, and a_{mx} terms for devices embedded in test fixtures, $\frac{a_m}{c_m}$, b_m, a_{mx}, and $e^{\gamma(L_{ref})}$ terms for on-wafer devices. On the other hand, the classical de-embedding method uses all element values of the matrices $\mathbf{T_x}$ and $\mathbf{T_y}$. Concerning the $\widetilde{S}_{12}$ and $\widetilde{S}_{21}$ parameters, it is observed that for coaxial devices and for devices embedded in a test fixture, only $\frac{a_m}{c_m}$ and b_m terms are required for its determination. That is, only partial knowledge of the $\mathbf{T_x}$ matrix is needed. In the special case of the on-wafer devices, the $\widetilde{S}_{12}$ and $\widetilde{S}_{21}$ parameters depend on $\frac{a_m}{c_m}$, b_m, and

a_{mx}, and $e^{\gamma(L_{\text{ref}})}$ terms. Procedures for computing $\frac{a_m}{c_m}$, b_m, a_{mx}, and $e^{\gamma(L_{\text{ref}})}$ are given next.

3.6 Computation of $\frac{a_m}{c_m}$, b_m using the L-L Method (Two Uniform Lines)

The L-L method [74] for computing the line propagation constant γ, b_m, and a_m/c_m, respectively, uses two uniform transmission lines of the different lengths referred to as L_{ref} and L_2 as shown in Figure 3.4. The only restriction is that one of them has to be longer than the other $(L_2 > L_{\text{ref}})$. The equivalent WCM M_2 resulting when L_2 is measured with the uncalibrated VNA system is given by,

$$\mathbf{M_2} = \mathbf{T_A}\,\mathbf{T_{L2}}\,\mathbf{T_B}. \tag{3.35}$$

In eqn (3.35), $\mathbf{T_{L2}}$ is the WCM of the uniform transmission line having Z_0 characteristic impedance, γ propagation constant, and mechanical length L_2 expressed as

$$\mathbf{T_{L2}} = \mathbf{T_\Gamma}\,\mathbf{T_{\lambda_2}}\,\mathbf{T_\Gamma}^{-1}, \tag{3.36}$$

In eqn (3.36), $\mathbf{T_\Gamma}$ is given by eqn (3.6) and $\mathbf{T_{\lambda_2}}$ by eqn 3.37

$$\mathbf{T_{\lambda_2}} = \begin{pmatrix} e^{-\gamma L_2} & 0 \\ 0 & e^{\gamma L_2} \end{pmatrix}. \tag{3.37}$$

Using eqn (3.18) and (3.35), the matrix product $\mathbf{M_1 M_2^{-1}}$ given by eqn (3.38) can be expressed as

$$\mathbf{M_{21}} = \mathbf{M_1 M_2^{-1}} = \mathbf{T_x T_{\lambda_x} T_x^{-1}}, \tag{3.38}$$

where

$$\mathbf{M_{21}} = \mathbf{M_1}\,\mathbf{M_2^{-1}} = \begin{pmatrix} t_{11} & t_{12} \\ t_{21} & t_{22} \end{pmatrix}. \tag{3.39}$$

Moreover, using eqn (3.20) and eqn (3.37), it is possible to express $\mathbf{T_{\lambda_x}}$ by

$$\mathbf{T_{\lambda_x}} = \mathbf{T_{\lambda_1}}\,\mathbf{T_{\lambda_2}^{-1}} = \begin{pmatrix} \lambda & \mathbf{0} \\ \mathbf{0} & \frac{1}{\lambda} \end{pmatrix}; \qquad \lambda = e^{\gamma(L_2 - L_{\text{ref}})}. \tag{3.40}$$

In eqn (3.38), $\mathbf{T_x}$ is given by eqn (3.16) and $\mathbf{T_{\lambda_x}}$ by eqn (3.40). Then, from eqn (3.38), it is possible to express $\mathbf{T_{\lambda_x}}$ as

$$\mathbf{T_{\lambda_x}} = \mathbf{T_x^{-1}}\,\mathbf{M_{21}}\,\mathbf{T_x}. \tag{3.41}$$

Now, substituting eqn (3.16)–(3.39) in the right-hand side of eqn (3.41) yields the following equation:

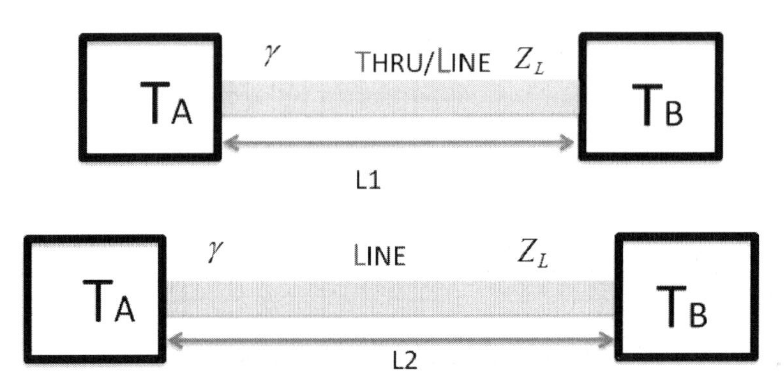

Figure 3.4 Structures used in the *two-line method.*

$$\begin{pmatrix} \lambda & 0 \\ 0 & \frac{1}{\lambda} \end{pmatrix} = \frac{1}{a_m - b_m\, c_m}$$

$$\begin{pmatrix} c_m[\frac{a_m}{c_m}\, t_{11} - \frac{a_m}{c_m}\, b_m\, t_{21} - b_m\, t_{22} + t_{12}] & -[{b_m}^2\, t_{21} + b_m\,(t_{22} - t_{11}) - t_{12}] \\ {c_m}^2(\frac{a_m}{c_m})^2 t_{21} + (\frac{a_m}{c_m})(t_{22} - t_{11}) - t_{12}] & c_m[\frac{a_m}{c_m}\, t_{22} - \frac{a_m}{c_m}\, b_m\, t_{21} - b_m\, t_{11} + t_{12}] \end{pmatrix} \tag{3.42}$$

$$\begin{pmatrix} \lambda & 0 \\ 0 & \frac{1}{\lambda} \end{pmatrix} = \frac{1}{a_m - b_m\, c_m} \begin{pmatrix} C_{11} & C_{12} \\ C_{21} & C_{22} \end{pmatrix} \tag{3.43}$$

$$B_{11} = c_m[\frac{a_m}{c_m}\, t_{11} - \frac{a_m}{c_m}\, b_m\, t_{21} - b_m\, t_{22} + t_{12}] \tag{3.44}$$

$$B_{12} = -[{b_m}^2\, t_{21} + b_m\,(t_{22} - t_{11}) - t_{12}] \tag{3.45}$$

$$B_{21} = {c_m}^2[(\frac{a_m}{c_m})^2 t_{21} + (\frac{a_m}{c_m})(t_{22} - t_{11}) - t_{12}] \tag{3.46}$$

$$B_{22} = c_m[\frac{a_m}{c_m}\, t_{22} - \frac{a_m}{c_m}\, b_m\, t_{f21} - b_m\, t_{11} + t_{12}]. \tag{3.47}$$

Comparing each term of the matrices on both sides of eqn (3.43), b_m, a_m/c_m, and λ are expressed as

$$b_m^2\, t_{21} + b_m\,(t_{22} - t_{11}) - t_{12} = 0, \tag{3.48}$$

$$\left(\frac{a_m}{c_m}\right)^2 t_{21} + \frac{a_m}{c_m}\,(t_{22} - t_{11}) - t_{12} = 0, \tag{3.49}$$

$$\lambda = \frac{\left[\frac{a_m}{c_m} t_{11} - \frac{a_m}{c_m} b_m t_{21} - b_m t_{22} + t_{12}\right]}{\frac{a_m}{c_m} - b_m}. \tag{3.50}$$

The quadratic eqn (3.48) and (3.49), first introduced in [21] and [74], are very important in the TRL calibration technique since they are used for the direct calculation of the calibration constant a_m/c_m, b_m and the term λ. The propagation constant γ is determined from eqn (3.50) as long as a_m/c_m and b_m are known

$$\gamma = \frac{1}{L_2 - L_{\text{ref}}} \ln \left\{ \frac{\left[\frac{a_m}{c_m} t_{11} - \frac{a_m}{c_m} b_m t_{21} - b_m t_{22} + t_{12}\right]}{\frac{a_m}{c_m} - b_m} \right\}. \tag{3.51}$$

It is important to mention that eqn (3.38) predicts the limitations of the L-L method at a frequency point where $\lambda = \pm 1$. In fact, the terms a_m/c_m, b_m are ill-conditioned when the argument of λ is $k\pi(k = 0, 1, 2, ..., n)$ since $\mathbf{M_1 M_2}^{-1}$ is the multiple of the identity matrix. Therefore, the S-parameters corrected with the TRL method exhibit discontinuities at these frequency points.

On the other hand, because equal coefficients appear in eqn (3.48) and (3.49), b_m and a_m/c_m are roots of the same equation. Moreover, since $\mathbf{T_A^{-1}}$ exists ($a_m - b_m c_m \neq 0$), then b_m must be different from a_m/c_m and hence b_m and a_m/c_m are the two different roots of eqn (3.48) and (3.49). Values of b_m and a_m/c_m are chosen according to the criterion reported on in [21] and [74], which states that

$$|b_m| < \left|\frac{a_m}{c_m}\right|. \tag{3.52}$$

Contrary to the classical methods using the wave propagation constant for the phase shift computation [51], eqn (3.50) gives straightforward information about the phase shift between the lines without previous knowledge of the wave propagation constant and, obviously, the physical dimensions of the lines. The expression (3.50) permits one to calculate, in the complex plane, the monotonous phase variations of λ in the whole frequency band.

3.7 Computation of a_m/c_m and b_m using the Line-Match (L-M) Method

The structures used for computing a_m/c_m and b_m with the line-match method are shown in Figure 3.5. The L-M method uses a reference line (the reference line can be a thru) and a match for determining b_m and a_m/c_m.

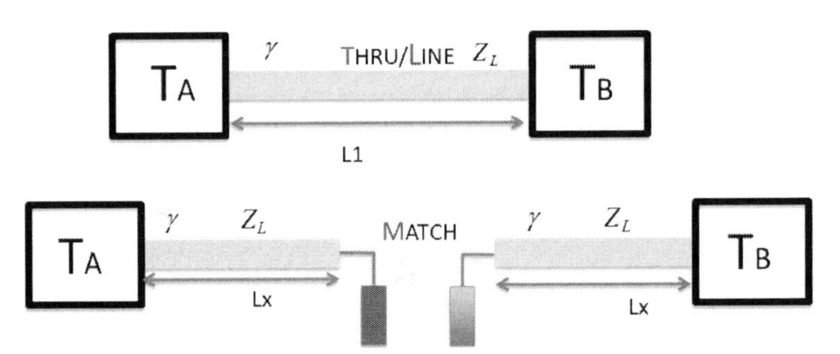

Figure 3.5 Structures used in the line-match method.

3.7.1 Line measurement

Using eqn (3.18) and (3.20), the equivalent matrix M_1 given by eqn (3.54) is written as

$$\mathbf{M_1} = \mathbf{T_x}\,\mathbf{T}_{\lambda_1}\mathbf{T_y}. \tag{3.53}$$

Then, solving for the $\mathbf{T_y}$ matrix, we have that

$$\mathbf{T_y} = \mathbf{T}_{\lambda_1}^{-1}\,\mathbf{T_x}^{-1}\mathbf{M_1}. \tag{3.54}$$

Here, $\mathbf{T_x}$, $\mathbf{T_y}$, and $\mathbf{T}_{\lambda_1}$ are expressed by eqn (3.16), (3.17), and (3.20) and $\mathbf{M_1}$ is defined as

$$\mathbf{M_1} = \begin{pmatrix} p_{11} & p_{12} \\ p_{21} & p_{22} \end{pmatrix}. \tag{3.55}$$

Developing eqn (3.54) and rearranging, terms α_m, β_m, ϕ_m, and $r_m\rho_m$ are given by

$$\alpha_m = \frac{1}{a_m}\left\{\frac{p_{11} - b_m p_{21}}{p_{22} - \frac{c_m}{a_m}p_{12}}\right\} e^{2\gamma L_{\mathrm{ref}}} \tag{3.56}$$

$$\beta_m = \frac{1}{a_m}\left\{\frac{p_{12} - b_m p_{22}}{p_{22} - \frac{c_m}{a_m}p_{12}}\right\} e^{2\gamma L_{\mathrm{ref}}} \tag{3.57}$$

$$\phi_m = \frac{\frac{a_m}{c_m}p_{21} - p_{11}}{\frac{a_m}{c_m}p_{22} - p_{12}} \tag{3.58}$$

$$r_m\alpha_m = \left\{\frac{\frac{a_m}{c_m}p_{22} - p_{12}}{\frac{a_m}{c_m} - b_m}\right\} e^{\gamma L_{\mathrm{ref}}}. \tag{3.59}$$

Solving for a_m/c_m from eqn (3.58), we have

$$\frac{a_m}{c_m} = \frac{p_{12}\phi_m - p_{11}}{\phi_m - p_{21}} \tag{3.60}$$

It follows from eqn (3.56) and (3.57) that β_m/α_m can be expressed as

$$\frac{\beta_m}{\alpha_m} = \frac{p_{12} - b_m\, p_{22}}{p_{11} - b_m\, p_{21}}. \tag{3.61}$$

Eqn (3.56)–(3.61) allow the calculation of α_m, β_m, ϕ_m, and $r_m\,\rho_m$, and β_m/α_m as long as a_m/c_m, b_m, and γ are known. The procedure to determine them is presented next.

3.7.2 Load measurements

It should be noted from eqn (3.60) that a_m/c_m depends on ϕ_m of the matrix $\mathbf{T_y}$ and p_{11}, p_{12}, and p_{21} of the matrix $\mathbf{M_1}$. The elements p_{11}, p_{12}, and p_{21} of $\mathbf{M_1}$ are all known elements. On the other hand, modeling the ports $\mathbf{T_A}$ and $\mathbf{T_B}$ by eqn (3.11) and (3.12), b_m and ϕ_m values can easily be determined as follows. The reflection coefficients S^{p1}_{11L} and S^{p2}_{22L}, as a function of the reflection coefficient Γ^{p1}_M, Γ^{p2}_M of the load and Γ^{p1}_L, Γ^{p2}_L of the line, of the structure shown in Figure 3.6 are derived for port 1 as

$$\mathbf{V_1} = \mathbf{T_A}\, \mathbf{T_{Lx1}}\, \mathbf{V_R}, \tag{3.62}$$

and for port 2 as

$$\mathbf{V_2} = \widetilde{\mathbf{T}}_{\mathbf{B}}^{-1}\, \mathbf{T_{Lx2}}\, \mathbf{V_R}, \tag{3.63}$$

with

$$\mathbf{T_{Lxi}} = \mathbf{T_\Gamma}\, \mathbf{T}_{\lambda_{\mathbf{xi}}}\, \mathbf{T_\Gamma}^{-1} \quad i = 1, 2. \tag{3.64}$$

In eqn (3.64), $\mathbf{T_\Gamma}$ is calculated with eqn (3.6), thus $\mathbf{T}_{\lambda_{\mathbf{xi}}}$ is given by

$$\mathbf{T}_{\lambda_{\mathbf{xi}}} = \begin{pmatrix} e^{-\gamma L_{xi}} & 0 \\ 0 & e^{\gamma L_{xi}} \end{pmatrix}, \; i = 1, 2. \tag{3.65}$$

In $\mathbf{T_{L_{xi}}}$, $\mathbf{T}_{\lambda_{\mathbf{xi}}}$ i=1 port 1, and i=2 port 2 and in eqn (3.62), $\mathbf{T_A}$ is given by eqn (3.11), $\mathbf{T_\Gamma}$ is given by eqn (3.6), $\widetilde{\mathbf{T}}_{\mathbf{B}} = \mathbf{X}\, \mathbf{T}_{\mathbf{B}}^{-1}\, \mathbf{X}$, $\mathbf{T}_{\mathbf{B}}^{-1}$, is the inverse matrix of $\mathbf{T_B}$ given by eqn (3.11) and $\mathbf{X}$, $\mathbf{V_1}$, $\mathbf{V_2}$,$\mathbf{V_R}$, are expressed as

$$\mathbf{X} = \begin{pmatrix} 0 & 1 \\ 1 & 0 \end{pmatrix}, \tag{3.66}$$

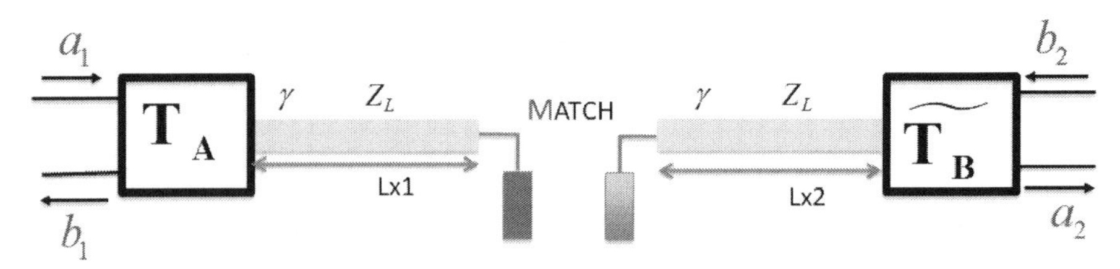

Figure 3.6 Match in port 1 and port 2.

$$\mathbf{V_1} = \begin{pmatrix} b_1 \\ a_1 \end{pmatrix}, \tag{3.67}$$

$$\mathbf{V_2} = \begin{pmatrix} b_2 \\ a_2 \end{pmatrix}, \tag{3.68}$$

$$\mathbf{V_R} = \begin{pmatrix} \Gamma_M^{p_i} \\ 1 \end{pmatrix} b_e. \tag{3.69}$$

Substituting eqn (3.64)–(3.69) into eqn (3.62)–(3.63) and defining the reflection coefficients $b_1/a_1 = S_{11L}^{p1}$ and $b_2/a_2 = S_{22L}^{p2}$, it follows that

$$S_{11L}^{p2} = \frac{a_m \lambda_{x1}^2 (\Gamma_M^{p1} - \Gamma_L^{p1}) + b_m (1 - \Gamma_M^{p1} \Gamma_L^{p1})}{c_m \lambda_{x1}^2 (\Gamma_M^{p1} - \Gamma_L^{p1}) + (1 - \Gamma_M^{p1} \Gamma_L^{p1})}, \tag{3.70}$$

and

$$S_{22L}^{p2} = \frac{\alpha_m \lambda_{x2}^2 (\Gamma_M^{p2} - \Gamma_L^{p2}) - \phi_m (1 - \Gamma_M^{p2} \Gamma_L^{p2})}{(1 - \Gamma_M^{p2} \Gamma_L^{p2}) - \beta_m \lambda_{x2}^2 (\Gamma_M^{p2} - \Gamma_L^{p2})}, \tag{3.71}$$

where

$$\lambda_{x1} = e^{\gamma L_{x1}} \tag{3.72}$$

$$\lambda_{x2} = e^{\gamma L_{x2}}. \tag{3.73}$$

It is clear that when $\Gamma_M^{p_i}$=$\Gamma_L^{p_i}$, eqn (3.70) and (3.71) become

$$b_m = S_{11L}^{p1} \tag{3.74}$$

$$\phi_m = -S_{22L}^{p2}. \tag{3.75}$$

This case occurs when the load and the line have identical impedance. Moreover, S_{11L}^{p1} and S_{22L}^{p2} are the scattering parameters measured when port 1 and port 2 are loaded with a broadband load.

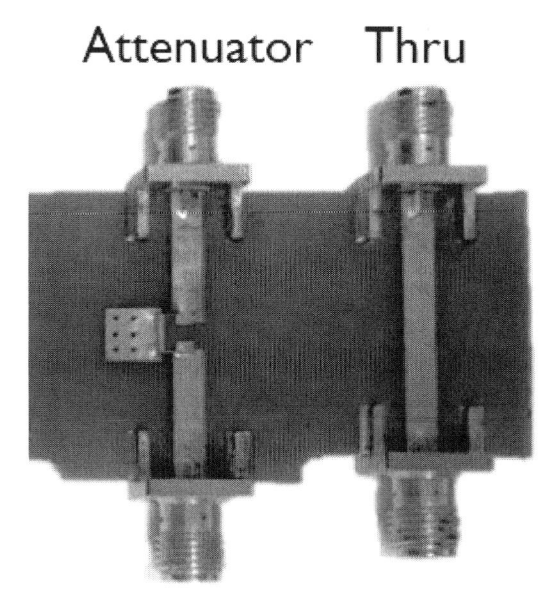

Figure 3.7 Calibration standards used in thru-attenuator (T-A) method.

3.8 Computation of a_m/c_m and b_m using the Thru-Attenuator (T-A) Method

The structure utilized for the determination of a_m/c_m and b_m using the thru-attenuator method is shown in Figure 3.7. Using the WCM formalism, a_m/c_m and b_m terms may be determined by using two different calibration elements: a thru and an attenuator. A symmetric attenuator may be modeled, using the WCM formalism as

$$\mathbf{A}_{\alpha} = \begin{pmatrix} A & 0 \\ 0 & \frac{1}{A} \end{pmatrix}; A = 10^{\frac{-\alpha}{20}}, \tag{3.76}$$

where α is the attenuation value in dB.

The resulting WCM $\mathbf{M_T}$ when the thru is measured with an uncalibrated VNA is given by

$$\mathbf{M_T} = \mathbf{T_x T_y}, \tag{3.77}$$

and the resulting WCM $\mathbf{M_A}$ when the attenuator is measured with an uncalibrated VNA is given by

$$\mathbf{M_A} = \mathbf{T_x A_{\alpha} T_y}. \tag{3.78}$$

Using eqn (3.77) and (3.78) and calculating the matrix product $\mathbf{M_A M_T^{-1}}$, the following expression results

$$\mathbf{M_A M_T^{-1}} = \mathbf{T_x A_\alpha T_x^{-1}}, \tag{3.79}$$

where

$$\mathbf{M_{AT}} = \mathbf{M_A M_T^{-1}}, \tag{3.80}$$

and expressed as

$$\mathbf{M_{AT}} = \begin{pmatrix} k_{11} & k_{12} \\ k_{21} & k_{22} \end{pmatrix}. \tag{3.81}$$

Using eqn (3.79) and solving for $\mathbf{A}_\alpha$, it follows that

$$\mathbf{A}_\alpha = \mathbf{T_x^{-1} M_A M_T^{-1} T_x}. \tag{3.82}$$

Now, substituting eqn (3.16), (3.81), and (3.76) in eqn (3.82) yields the following equation:

$$\begin{pmatrix} A & 0 \\ 0 & \frac{1}{A} \end{pmatrix} = \frac{1}{a_m - b_m\, c_m} \begin{pmatrix} C_{11} & C_{12} \\ C_{21} & C_{22} \end{pmatrix}, \tag{3.83}$$

where the matrix elements C_{11}, C_{12}, C_{21}, and C_{22} are expressed as

$$C_{11} = c_m[\frac{a_m}{c_m}\, k_{11} - \frac{a_m}{c_m}\, b_m\, k_{f21} - b_m\, k_{22} + k_{12}] \tag{3.84}$$

$$C_{12} = -[b_m{}^2\, k_{21} + b_m\, (k_{22} - k_{11}) - k_{12}] \tag{3.85}$$

$$C_{21} = c_m{}^2[(\frac{a_m}{c_m})^2 k_{21} + (\frac{a_m}{c_m})(k_{22} - k_{11}) - k_{12}] \tag{3.86}$$

$$C_{22} = c_m[\frac{a_m}{c_m}\, k_{22} - \frac{a_m}{c_m}\, b_m\, k_{21} - b_m\, k_{11} + k_{12}]. \tag{3.87}$$

Next, comparing the off-diagonal terms of the left-hand side with the off-diagonal terms of the right-hand side of eqn (2.82) expressed by eqn (2.76) and eqn (2.83), the following quadratic equation for computing a_m/c_m and b_m results:

$$b_m^2\, k_{21} + b_m\, (k_{22} - k_{11}) - k_{12} = 0, \tag{3.88}$$

$$\left(\frac{a_m}{c_m}\right)^2\, k_{21} + \frac{a_m}{c_m}\, (k_{22} - k_{11}) - k_{12} = 0. \tag{3.89}$$

Comparing each term of the main diagonal eqn (3.76) of matrices on both sides of eqn (3.83) A is expressed as

$$A = \frac{[\frac{a_m}{c_m} k_{11} - \frac{a_m}{c_m} b_m k_{21} - b_m k_{22} + k_{12}]}{\frac{a_m}{c_m} - b_m}. \tag{3.90}$$

3.9 Calculation of a_m

The structures utilized in a_m calculation are shown in Figure 3.8.

$$S_{11R}^{p2} = \frac{a_m \lambda_{x1}^2 (\Gamma_R^{p1} - \Gamma_L^{p1}) + b_m (1 - \Gamma_R^{p1} \Gamma_L^{p1})}{c_m \lambda_{x1}^2 (\Gamma_R^{p1} - \Gamma_L^{p1}) + (1 - \Gamma_R^{p1} \Gamma_L^{p1})}, \tag{3.91}$$

and

$$S_{22RL}^{p2} = \frac{\alpha_m \lambda_{x2}^2 (\Gamma_R^{p2} - \Gamma_L^{p2}) - \phi_m (1 - \Gamma_R^{p2} \Gamma_L^{p2})}{(1 - \Gamma_R^{p2} \Gamma_L^{p2}) - \beta_m \lambda_{x2}^2 (\Gamma_R^{p2} - \Gamma_L^{p2})}, \tag{3.92}$$

where Γ_R^{p1}, Γ_R^{p2} are the load coefficient at the end of the line connected at port 1 and port 2, respectively, as shown in Figure 3.8. Solving for Γ_R^{p1}, Γ_R^{p2} from eqn (3.91) and (3.92), one has

$$\Gamma_R^{p1} = \frac{(b_m - S_{11R}^{p1}) + a_m \lambda_{x1}^2 (S_{11R}^{p1} \frac{c_m}{a_m} - 1) \Gamma_L^{p1}}{a_m \lambda_{x1}^2 (S_{11R}^{p1} \frac{c_m}{a_m} - 1) + (b_m - S_{11R}^{p1}) \Gamma_L^{p1}}, \tag{3.93}$$

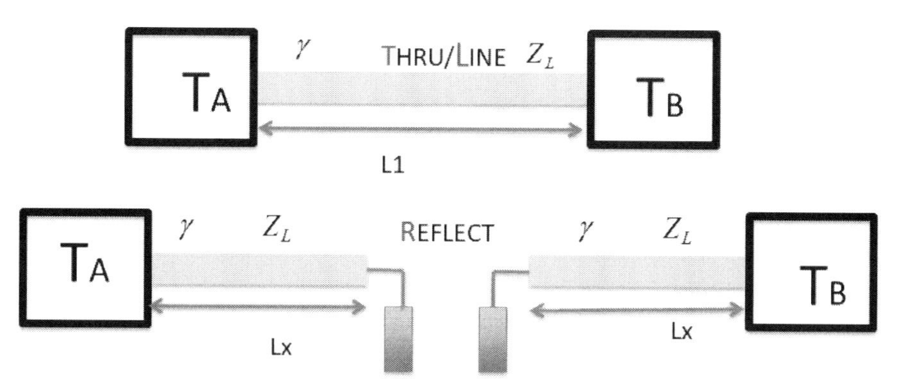

Figure 3.8 Structures utilized for the derivation of a_m.

and

$$\Gamma_R^{p_2} = \frac{(\phi_m + S_{22R}^{p_2}) + \lambda_{x2}^2 \Gamma_L^{p_2}(\alpha_m + S_{22R}^{p_2}\beta_m)}{\Gamma_L^{p_2}(\phi_m + S_{22R}^{p_2}) + \lambda_{x2}^2(\alpha_m + S_{22R}^{p_2}\beta_m)}. \tag{3.94}$$

The thru/line-reflect-line, (TRL/LRL), the thru/line-reflect-match, (TRM/LRM), the thru/line-reflect-attenuator, (TRA/LRA) require that $\Gamma_R^{p_1} = \Gamma_R^{p_2}$. This hypothesis and the utilization of eqn (3.56) calculated with the reference line leads to the following result:

$$a_m = \frac{\lambda_{x2}}{\lambda_{x1}} \pm \left\{ \sqrt{\frac{(p_{11} - b_m p_{21})(b_m - S_{11R}^{p_1})(1 - \frac{\beta_m}{\alpha_m} S_{22R}^{p_2})}{(p_{22} - \frac{c_m}{a_m} p_{12})(\phi_m + S_{22R}^{p_2})(\frac{c_m}{a_m} S_{11R}^{p_1} - 1)}} \right\} e^{\gamma L_{\text{ref}}}. \tag{3.95}$$

Thus, the calculation of a_m using eqn (3.95) requires previous knowledge of $e^{\gamma L_{\text{ref}}}$. However, when $\lambda_{x1} = \lambda_{x2}$ and the length of reference line L_{ref} is zero, as the case of TRL, TRM, and TAR, the calculation of a_m is direct and simple. The calculation of a_m is more complicated when $\lambda_{x1} \neq \lambda_{x2}$ and the length of the reference line L_{ref} is arbitrary.

4

VNA Calibration Techniques

4.1 Introduction

A calibrated vector network analyzer must be used to measure the small-signal behavior of a device under test (DUT) at high frequencies, by means of S-parameters [44]. The reliability of the measured S-parameter depends on the accuracy of its calibration [80]. This chapter presents the most important VNA calibration techniques: thru-reflect-line (TRL), thru-reflect-match (TRM), thru-reflect-reflect-match (TRRM), and short-open-load-thru/reciprocal (SOLT), along with the line-offset open-offset short (LZZ) calibration.

The mathematical procedures presented in this chapter are based on the use of ABCD-parameters, which are not defined as a function of a reference impedance, contrary to the T-parameters that are typically used in the development of calibration procedures. Calibration procedures developed using T-parameters refer the calibration to the (frequency-dependent) impedance of one of the calibration structures used in a calibration method: typically a load or a transmission line. Then, a transformation of reference impedance from the native frequency-dependent value to the frequency-independent impedance of the measurement system Z_0 is required. The mathematical formalism used in this chapter shows that, as long as the impedance of the loads used in TRM or TRRM calibrations or the characteristic impedance of the lines is used in TRL, LZZ, or LZZM, the unique impedance to which the calibration is referred is the measurement system impedance.

4.2 ABCD-Parameters Representation of a Transmission Line

Two-port calibration techniques use the connection of the VNA ports at the calibration plane as calibration element. Such a connection is typically achieved by using a uniform transmission line.[1] Even the cases in which this connection is achieved by using a through connection of the calibration reference planes (thru) may be considered as a zero-length transmission line [65]. The electrical behavior of a uniform transmission line may be described by its characteristic impedance Z_L, propagation constant γ, and length l. The ABCD-parameters representation of a uniform transmission line $\mathbf{T_L}$ is denoted as

$$\mathbf{T_L} = \begin{bmatrix} \cosh(\gamma l) & Z_L \sinh(\gamma l) \\ Z_L^{-1} \sinh(\gamma l) & \cosh(\gamma l) \end{bmatrix}. \tag{4.1}$$

In the calibration procedures described in this section, a powerful tool of linear algebra, matrix diagonalization, is applied to the analysis of the electrical properties of a uniform transmission line. Matrix diagonalization is used to decompose the ABCD-parameters matrix of a transmission line into a matrix product, with important implications in the analysis of VNA calibration techniques. As described in Appendix A, it is possible to express the matrix $\mathbf{T_L}$ as a product of the form

$$\mathbf{T_L} = \mathbf{T_Z} \mathbf{T}_\lambda \mathbf{T_Z}^{-1}, \tag{4.2}$$

where

$$\mathbf{T}_\lambda = \begin{bmatrix} \lambda_L & 0 \\ 0 & \lambda_L^{-1} \end{bmatrix} \tag{4.3}$$

and

$$\mathbf{T_Z} = \begin{bmatrix} -Z_L & Z_L \\ 1 & 1 \end{bmatrix}. \tag{4.4}$$

In eqn (4.3), $\lambda_L = e^{-\gamma l}$, $\gamma = \alpha + j\beta$, with α and β being the per-unit-length losses and phase constants of a transmission line. A generalized approach is summarized in Table 4.1, in which different forms of diagonalization of $\mathbf{T_L}$, depending on the definition of the matrix of eigenvalues ($\mathbf{T}_\lambda$), are presented. Table 4.1 also shows the diagonalization of the T-parameters matrix of a transmission line, which is used in Chapter 3 to describe the microwave de-embedding method.

[1]The short-open-load-reciprocal (SOLR) [27] is the only technique that allows using an arbitrary reciprocal two-port device as calibration element, not strictly a uniform transmission line.

Table 4.1 Forms of diagonalization of the T- and ABCD-parameters of a uniform transmission line.

	Diagonalization 1	Diagonalization 2
Matrix of eigenvalues	$\mathbf{T}_\lambda = \begin{bmatrix} \lambda_L & 0 \\ 0 & \lambda_L^{-1} \end{bmatrix}$	$\mathbf{T}_\lambda = \begin{bmatrix} \lambda_L^{-1} & 0 \\ 0 & \lambda_L \end{bmatrix}$
Matrix of eigenvectors in T-parameters	$\mathbf{T_Z} = \begin{bmatrix} 1 & \Gamma_L \\ \Gamma_L & 1 \end{bmatrix}$	$\mathbf{T_Z} = \begin{bmatrix} \Gamma_L & 1 \\ 1 & \Gamma_L \end{bmatrix}$
Matrix of eigenvectors in ABCD-parameters	$\mathbf{T_Z} = \begin{bmatrix} -Z_L & Z_L \\ 1 & 1 \end{bmatrix}$	$\mathbf{T_Z} = \begin{bmatrix} Z_L & -1 \\ 1 & Z_L^{-1} \end{bmatrix}$

The diagonalization of the ABCD-parameters matrix representation of a transmission line is very useful in the development of calibration algorithms for the TRL technique allowing the use of lines of arbitrary impedance as calibration structures. In the case of a zero-length thru, $\mathbf{T}_\lambda$ is identical to the 2×2 identity matrix, $\mathbf{I_2}$. In such a case, $\mathbf{T_L} = \mathbf{I_2} = \mathbf{T_Z}\mathbf{T_Z}^{-1}$, independently of the definition of the matrix of eigenvectors ($\mathbf{T_Z}$). Thus, in a more general case, the matrix $\mathbf{T_Z}$ may be defined as

$$\mathbf{T_Z} = \begin{bmatrix} -Z_B & Z_A \\ 1 & 1 \end{bmatrix}. \tag{4.5}$$

When using a thru instead of a line as calibration element, the identity $\mathbf{I_2} = \mathbf{T_Z}\mathbf{T_Z}^{-1}$ holds even though Z_A and Z_B are different from Z_L. This result is useful to develop calibration algorithms for the TRM technique allowing the use of loads of arbitrary impedance, and the TRL technique using lines of arbitrary impedance.

4.3 Measurement of Calibration Structures

The calibration procedures described in this section are based on the use of the eight-term error model; this means that the isolation and switching errors are either ignored or have been corrected in a step different from the calibration. When using the eight-term error model and the ABCD-parameters, the measurement of a DUT (Figure 4.1) can be mathematically expressed as

$$\mathbf{M_D} = \mathbf{T_A}\mathbf{T_D}\mathbf{T_B}. \tag{4.6}$$

In eqn (4.6), $\mathbf{M_D}$ and $\mathbf{T_D}$ are ABCD-parameters matrices representing the uncorrected and the actual behavior of a DUT, respectively. Meanwhile, $\mathbf{T_A}$ and $\mathbf{T_B}$ are matrices representing the errors in ports 1 and 2 of the VNA,

Figure 4.1 Measurement of a two-port DUT. Ref [71]. © 2012 IEEE. Reprinted with permission.

respectively. The measurement of a DUT using a VNA may be then depicted as shown in Figure 4.1.

The problem to solve in calibration is to calculate the elements of the matrices $\mathbf{T_A}$ and $\mathbf{T_B}$, or some modified version of them, using information provided by the measurement of a determined set of calibration structures. Once these terms are known, the ABCD-parameters of a DUT may be determined by solving eqn (4.6) for $\mathbf{T_D}$. In this section, the mathematical analysis corresponding to the measurement of different calibration structures used in TRL, TRM, TRRM, and LZZ calibration techniques is presented. The information provided by the measurement of these structures is used in the following section to explain how different calibration techniques determine the calibration error terms.

4.3.1 Measurement of a transmission line

Since the ABCD-parameters matrix of a uniform transmission line may be represented by the matrix product $\mathbf{T_L} = \mathbf{T_Z}\mathbf{T_\lambda}\mathbf{T_Z^{-1}}$, the measurement of a transmission line may be expressed as

$$\mathbf{M_L} = \mathbf{T_A}\mathbf{T_L}\mathbf{T_B} = \mathbf{T_X}\mathbf{T_\lambda}\mathbf{T_Y} = \begin{bmatrix} p_{11} & p_{12} \\ p_{21} & p_{22} \end{bmatrix}, \tag{4.7}$$

where

$$\mathbf{T_X} = \mathbf{T_A}\mathbf{T_Z} \triangleq D_X\overline{\mathbf{T_X}} = D_X \begin{bmatrix} \overline{A_X} & \overline{B_X} \\ \overline{C_X} & 1 \end{bmatrix}, \tag{4.8}$$

$$\mathbf{T_Y} = \mathbf{T_Z^{-1}}\mathbf{T_B} \triangleq D_Y\overline{\mathbf{T_Y}} = D_Y \begin{bmatrix} \overline{A_Y} & \overline{B_Y} \\ \overline{C_Y} & 1 \end{bmatrix}. \tag{4.9}$$

Solving eqn (4.6) for $\mathbf{T_D}$ and using eqn (4.8) and (4.9) in the resulting expression, $\mathbf{T_D}$ may be expressed as

$$\mathbf{T_D} = \mathbf{T_Z}\mathbf{T_X^{-1}}\mathbf{M_D}\mathbf{T_Y^{-1}}\mathbf{T_Z^{-1}}. \tag{4.10}$$

In order to determine the ABCD-parameters of a DUT, provided that the matrix $\mathbf{T_Z}$ is known, seven terms have to be calculated during the calibration process: three elements of the matrix $\mathbf{T_X}$, namely $\overline{A_X}$, $\overline{B_X}$, $\overline{C_X}$, three elements of the matrix $\mathbf{T_Y}$, namely $\overline{A_Y}$, $\overline{B_Y}$, $\overline{C_Y}$, and the product $D_X D_Y$.

Using the measurement of a transmission line, it is possible to determine expressions for four of the seven calibration terms. Solving eqn (4.7) for $\mathbf{T_Y}$ and using eqn (4.8) and (4.9), the following expressions may be obtained:

$$\overline{A_Y} = \frac{1}{\overline{C_X}} \frac{p_{11} - \overline{B_X} p_{21}}{-p_{12} + \frac{\overline{A_X}}{\overline{C_X}} p_{22}} \lambda_L^{-2}, \tag{4.11}$$

$$\overline{B_Y} = \frac{1}{\overline{C_X}} \frac{p_{12} - \overline{B_X} p_{22}}{-p_{12} + \frac{\overline{A_X}}{\overline{C_X}} p_{22}} \lambda_L^{-2}, \tag{4.12}$$

$$\overline{C_Y} = \frac{-p_{11} + \frac{\overline{A_X}}{\overline{C_X}} p_{21}}{-p_{12} + \frac{\overline{A_X}}{\overline{C_X}} p_{22}}, \tag{4.13}$$

$$D_X D_Y = \frac{\frac{\overline{A_X}}{\overline{C_X}} p_{22} - p_{12}}{\frac{\overline{A_X}}{\overline{C_X}} - \overline{B_X}} \lambda_L. \tag{4.14}$$

Note that for calculating these terms, $\frac{\overline{A_X}}{\overline{C_X}}$, $\overline{B_X}$, and $\overline{C_X}$ have to be known. The following sections show how to determine $\frac{\overline{A_X}}{\overline{C_X}}$, $\overline{B_X}$, and $\overline{C_X}$ by combining the measurements of different sets of calibration structures.

Measurement of a thru:

A special case of the measurement of the connection of the VNA ports is achieved by using the through connection of the VNA ports (*thru*). A thru structure may be mathematically analyzed as a zero-length transmission line. Let $\mathbf{T_T}$, defined as

$$\mathbf{T_T} = \mathbf{I_2} = \mathbf{T_Z} \mathbf{T_Z^{-1}}, \tag{4.15}$$

be an ABCD-parameters matrix denoting the through connection of the VNA ports. Then, the measurement of a thru structure may be represented as

$$\mathbf{M_T} = \mathbf{T_A} \mathbf{T_T} \mathbf{T_B}. \tag{4.16}$$

By substituting eqn (4.15) in eqn (4.16), it is possible to express $\mathbf{M_T}$ as

$$\mathbf{M_T} = \mathbf{T_X}\mathbf{T_Y}. \tag{4.17}$$

Thus, in the case of a zero-length thru ($l = 0$), the terms $\overline{A_Y}$, $\overline{B_Y}$, $\overline{C_Y}$, and $D_X D_Y$ can be calculated from eqn (4.11) and (4.14) by using $\lambda_L = 1$.

4.3.2 Measurement of a pair of loads

VNA calibration techniques use the measurement of at least one pair of loads that must be connected at both ports of the VNA. The most commonly used loads are short circuits, open circuits, and loads of impedance value close to Z_0 (*match*). In the analysis presented in the following sections O, S, and M will denote the use of open, short, and match standards, respectively. General expressions for the impedance at the input of port $P = 1, 2$ of the VNA, Z^m_{KP} ($K = O, S, M$), when it is loaded with a load of impedance Z_{KP} are obtained and the information provided by the measurement of these loads are studied.

The ABCD-parameters matrix of a shunted load of impedance Z_{KP} (Figure 4.2) is expressed as [18]

$$\mathbf{T_{KP}} = \begin{bmatrix} 1 & 0 \\ Z_{KP}^{-1} & 1 \end{bmatrix}. \tag{4.18}$$

The loads used as calibration elements in VNA calibration techniques are one-port terminations. Thus, it is possible to analyze these structures by using the concept of *impedance at the input of a two-port network under output open-circuit condition*.

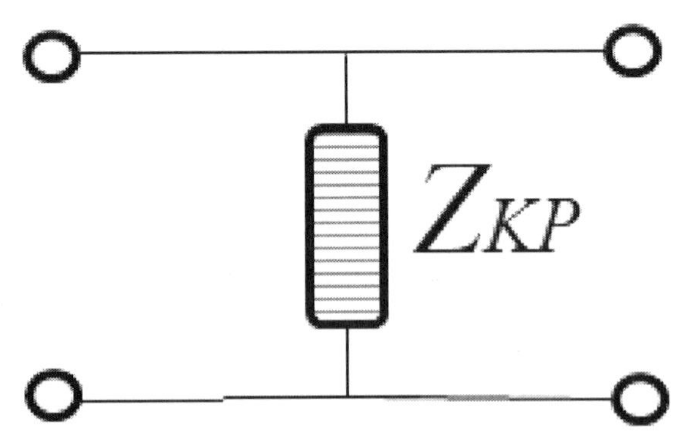

Figure 4.2 Equivalent two-port structure of a shunted load. Ref [62]. © 2013 IEEE. Reprinted with permission.

Measurement of a load at port 1:
The equivalent ABCD-parameters matrix of the structure enclosed in dotted lines in Figure 4.3(a), $\mathbf{M_{K1}}$, may be expressed as

$$\mathbf{M_{K1}} = \mathbf{T_A T_{K1}}. \tag{4.19}$$

By solving eqn (4.8) for $\mathbf{T_A}$ and substituting the resulting expression in eqn (4.19), $\mathbf{M_{K1}}$ may be expressed as

$$\mathbf{M_{K1}} = \mathbf{T_X T_Z^{-1} T_{K1}}. \tag{4.20}$$

Then, by developing eqn (4.20) and calculating the impedance at the input of the structure shown in Figure 4.3(a) under output open-circuit condition, one has

$$Z_{K1}^{m} = \frac{\overline{A_X}\left(Z_{K1} - Z_A\right) - \overline{B_X}\left(Z_{K1} + Z_B\right)}{\overline{C_X}\left(Z_{K1} - Z_A\right) - \left(Z_{K1} + Z_B\right)}. \tag{4.21}$$

Measurement of a load at port 2:
The equivalent ABCD-parameters matrix of the structure shown in Figure 4.3(b), $\mathbf{M_{K2}}$, may be expressed as

$$\mathbf{M_{K2}} = \widetilde{\mathbf{T_B}}\mathbf{T_{K2}}. \tag{4.22}$$

$\widetilde{\mathbf{T_B}}$ is the reverse cascading ABCD-parameters matrix corresponding to $\mathbf{T_B}$, which is defined as

$$\widetilde{\mathbf{T_B}} = \mathbf{X T_B^{-1} X}, \tag{4.23}$$

where

$$\mathbf{X} = \begin{bmatrix} \pm 1 & 0 \\ 0 & \mp 1 \end{bmatrix}. \tag{4.24}$$

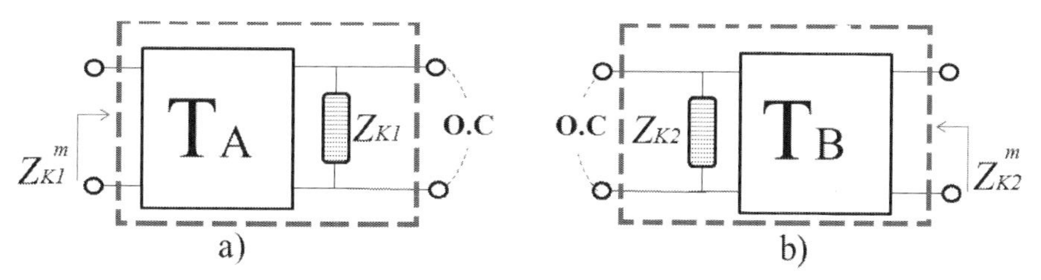

Figure 4.3 Equivalent structures of the measurement of loads connected at: (a) port 1 y and (b) port 2 of the VNA. Ref [62]. © 2013 IEEE. Reprinted with permission.

By substituting eqn (4.23) and (4.24) in eqn (4.22), $\mathbf{M_{K2}}$ may be expressed as

$$\mathbf{M_{K2}} = \mathbf{X T_B^{-1} X T_{K2}}. \tag{4.25}$$

By solving eqn (4.9) for $\mathbf{T_B}$ and substituting the resultant expression in eqn (4.25), $\mathbf{M_{K2}}$ may be expressed as

$$\mathbf{M_{K2}} = \mathbf{X T_Y^{-1} T_Z^{-1} X T_{K2}}. \tag{4.26}$$

Finally, by developing eqn (4.26) and calculating the impedance at the input of the structure shown in Figure 4.3(b) under output open-circuit condition, one has

$$Z_{K2}^{m} = \frac{\overline{B_Y}\left(Z_{K2} - Z_B\right) + \left(Z_{K2} + Z_A\right)}{\overline{A_Y}\left(Z_{K2} - Z_B\right) + \overline{C_Y}\left(Z_{K2} + Z_A\right)}. \tag{4.27}$$

Measurement of a pair of reflecting loads:

VNA calibration techniques use at least one pair of symmetrical reflecting loads connected at ports 1 and 2 of the VNA[2], for determining the $\overline{C_X}$ term.

The loads commonly used as reflecting loads are either short circuits or open circuits. The subindex *K* in eqn (4.21) and (4.27) is substituted here by *R* = *O, S* in order to denote the use of a highly reflecting load. Regarding VNA port 1, by solving eqn (4.21) for Z_{K1} ($\equiv Z_{R1}$), one has

$$Z_{R1} = \frac{Z_A \overline{C_X} + Z_B \eta_{R1}}{\overline{C_X} - \eta_{R1}}, \tag{4.28}$$

where

$$\eta_{R1} = \frac{Z_{R1}^{m} - \overline{B_X}}{Z_{R1}^{m} - \frac{\overline{A_X}}{\overline{C_X}}}. \tag{4.29}$$

Similarly, at port 2 of the VNA, by solving eqn (4.27) for Z_{K2} ($\equiv Z_{R2}$), one has

$$Z_{R2} = \frac{Z_B \overline{A_Y} + Z_A \eta_{R2}}{\overline{A_Y} - \eta_{R2}}, \tag{4.30}$$

[2]Symmetry recalls to the condition in which the load connected at port 1 is identical to the load connected at port 2.

where

$$\eta_{R2} = \frac{1 - Z_{R2}^{m}\overline{C_Y}}{Z_{R2}^{m} - \frac{\overline{B_Y}}{\overline{A_Y}}}. \tag{4.31}$$

Since the reflecting loads are assumed symmetrical, $Z_{R1} = Z_{R2}$, by equating eqn (4.28) to (4.30), the following equation for the term $\overline{C_X}$ may be obtained:

$$\overline{C_X} = \frac{-c_1 \pm \sqrt{c_1^2 - 4c_2c_2}}{2c_2}, \tag{4.32}$$

where

$$c_0 = -Z_B\eta_{R1}\left(\overline{A_Y} \cdot \overline{C_X}\right), \tag{4.33}$$

$$c_1 = \frac{1}{2}\left(\overline{A_Y} \cdot \overline{C_X} + \eta_{R1}\eta_2\right)\left(Z_B - Z_A\right), \tag{4.34}$$

$$c_2 = Z_A\eta_{R2}. \tag{4.35}$$

In most of the calibration techniques $Z_A = Z_B$, and then eqn (4.32) reduces to

$$\overline{C_X} = \pm\sqrt{\left(\overline{A_Y} \cdot \overline{C_X}\right)\eta_{R1}/\eta_{R2}}. \tag{4.36}$$

For calculating $\overline{C_X}$ it is necessary to determine $\frac{\overline{A_X}}{\overline{C_X}}$, and $\overline{B_X}$. In the following sections, it is described how to determine these terms by using the calibration structures corresponding to different calibration techniques.

4.4 The TRL Calibration Technique using Lines of Arbitrary Impedance

The thru-reflect-line (TRL) calibration technique uses as calibration elements the through connection of the VNA ports, a transmission line of characteristic impedance Z_L and a pair of loads of impedance Z_{R1}, $Z_{R2} \neq Z_L$ connected at both ports of the VNA[3], as shown in Figure 4.4.

The TRL calibration technique was introduced by G. Engen in [21] using T-parameters to represent the VNA and the calibration elements; it assumes that the line used as calibration element is a non-reflecting line, implying

[3]The TRL can also be implemented by using two lines of different lengths and of identical characteristic impedance and propagation constant. This method is known as LRL (line-reflect-line).

that the calibration reference impedance is identical to Z_L. Then, in [53], a procedure for transforming the calibration reference impedance of the TRL from Z_L to Z_0 after the calibration was introduced. The TRL procedure described here refers the calibration to Z_0, without the use of post-calibration impedance transformations.

The TRL calibration technique determines the $\overline{\frac{A_X}{C_X}}$ and $\overline{B_X}$ terms using the thru-line pair combination. The ABCD-parameters matrix of the thru structure shown in Figure 4.4(a) may be denoted as $\mathbf{M_T} = \mathbf{T_X}\mathbf{T_Y}$. Meanwhile, the ABCD-parameters matrix of the line structure shown in Figure 4.4(b) may be denoted as $\mathbf{M_L} = \mathbf{T_X}\mathbf{T_\lambda}\mathbf{T_Y}$. Then, by calculating the matrix product $\mathbf{M_{LT}} = \mathbf{M_L}\mathbf{M_T^{-1}}$, one has

$$\mathbf{M_{LT}} = \mathbf{T_X}\mathbf{T_\lambda}\mathbf{T_X^{-1}} = \begin{bmatrix} m_{11} & m_{12} \\ m_{21} & m_{22} \end{bmatrix}. \tag{4.37}$$

Then, by substituting eqn (4.2) and (4.8) in eqn (4.37) and developing the resulting expression, the following equations may be derived:

$$\overline{\frac{A_X}{C_X}}, \overline{B_X} = \frac{(m_{11} - m_{22}) \pm \sqrt{(m_{11} - m_{22})^2 - 4(m_{12}m_{21})}}{2m_{21}}, \tag{4.38}$$

$$\lambda_L = \frac{\overline{\frac{A_X}{C_X}} m_{11} + m_{21} - \overline{\frac{A_X}{C_X}}\,\overline{B_X} m_{21} - \overline{B_X} m_{22}}{\overline{\frac{A_X}{C_X}} - \overline{B_X}}. \tag{4.39}$$

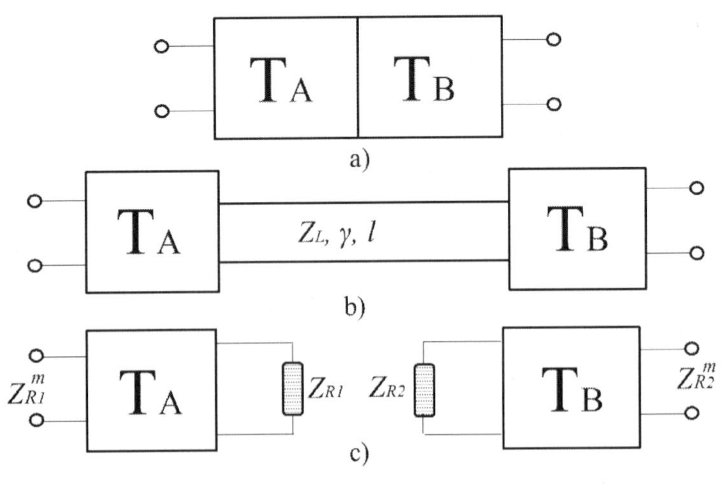

Figure 4.4 TRL calibration structures: (a) thru, (b) line, and (c) reflecting load. Ref [72]. © 2013 IEEE. Reprinted with permission.

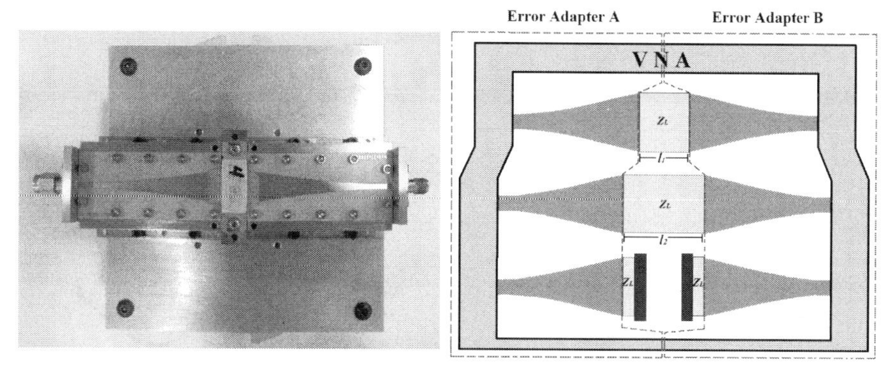

Figure 4.5 Test fixture using Klopfestain impedance adapter (left) and an illustration of the corresponding TRL calibration structures (right). Ref [72]. © 2013 IEEE. Reprinted with permission.

In eqn (4.38), one of the roots corresponds to $\overline{\frac{A_X}{C_X}}$ and the other root corresponds to $\overline{B_X}$. The values of $\overline{\frac{A_X}{C_X}}$ and $\overline{B_X}$ have to be chosen such that they allow calculating a continuous value of λ using eqn (4.39) [74, 71]. Eqn (4.39) allows to determine the wave vector, from which the transmission line propagation constant may be determined. The ABCD-parameters of a DUT may be determined by using eqn (4.10), where the characteristic impedance of the transmission line used in the TRL is included through the matrix $\mathbf{T_Z}$, thus allowing the use of lines of arbitrary impedance. Figure 4.5 shows an example of a test fixture aimed to accommodate high power transistors, including low impedance transmission lines. TRL calibration can be implemented by tailoring calibration structures (lines and reflecting loads) in such test fixture [72]. In Figure 4.6, the S-parameters of a high power transistor are corrected using the classical TRL procedure (using T-parameters and reference impedance transformation) and the procedure described in this section (using ABCD-parameters and avoiding impedance transformation).

Limitations of the TRL technique:

The frequency bandwidth of the measurements corrected using the TRL is limited since the equations characterizing the TRL are ill-conditioned at frequencies at which the phase shift of the thru-line pair combination is $k \cdot 180°$, $k \in \mathbb{N}$. At these frequencies, the matrices $\mathbf{T_X}$ and $\mathbf{T_Y}$ are non-singular matrices. The matrix $\mathbf{T_X}$, for example, is non-singular if its determinant, equals zero (i.e., $\overline{\frac{A_X}{C_X}} = \overline{B_X}$). By multiplying both sides of

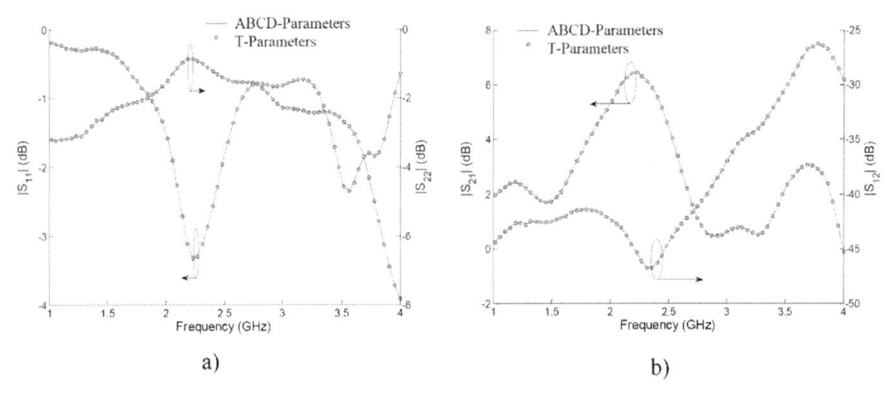

Figure 4.6 S-parameters: (a) S_{11}, S_{22} and (b) S_{12}, S_{21}, corrected using the classical TRL procedure and the procedure implemented using ABCD-parameters. Ref [72]. © 2013 IEEE. Reprinted with permission.

eqn (4.37), by their right-hand side by $\mathbf{T_X}$, and equating term by term the resultant equations, the following expressions may be obtained:

$$\det(\mathbf{T_X}) = \overline{A_X} - \overline{B_X C_X} = \overline{C_X}\left(\frac{\overline{A_X}}{\overline{C_X}} - \overline{B_X}\right), \tag{4.40}$$

$$\frac{\overline{A_X}}{\overline{C_X}} = \frac{m_{12}}{\lambda - m_{11}}, \tag{4.41}$$

$$\overline{B_X} = \frac{m_{12}}{\lambda^{-1} - m_{11}}. \tag{4.42}$$

From eqn (4.41) and (4.42), it can be noted that $\frac{\overline{A_X}}{\overline{C_X}} = \overline{B_X}$ when $\lambda = e^{\gamma l} = 1$, which occurs when βl is a multiple of $180°$ and αl is close to zero. Hence, a very short transmission line is useful to avoid βl being close to a multiple of $180°$ at high frequencies; meanwhile, long transmission lines are useful to avoid βl being close to $0°$ at low frequencies. The use of different lines for covering a wide frequency bandwidth is known in the literature as the multi-line TRL (mTRL) technique [51].

The Figure 4.7, shows an example of calibration results of a DUT (offset-short) that may be obtained when using lines whose phase difference presents multiple crossings at $k \cdot 180°$. TRL calibration structures are achieved by using rectangular waveguides in the X-band. In this example, two lines of different lengths and a symmetrical short circuit are used as TRL structures.

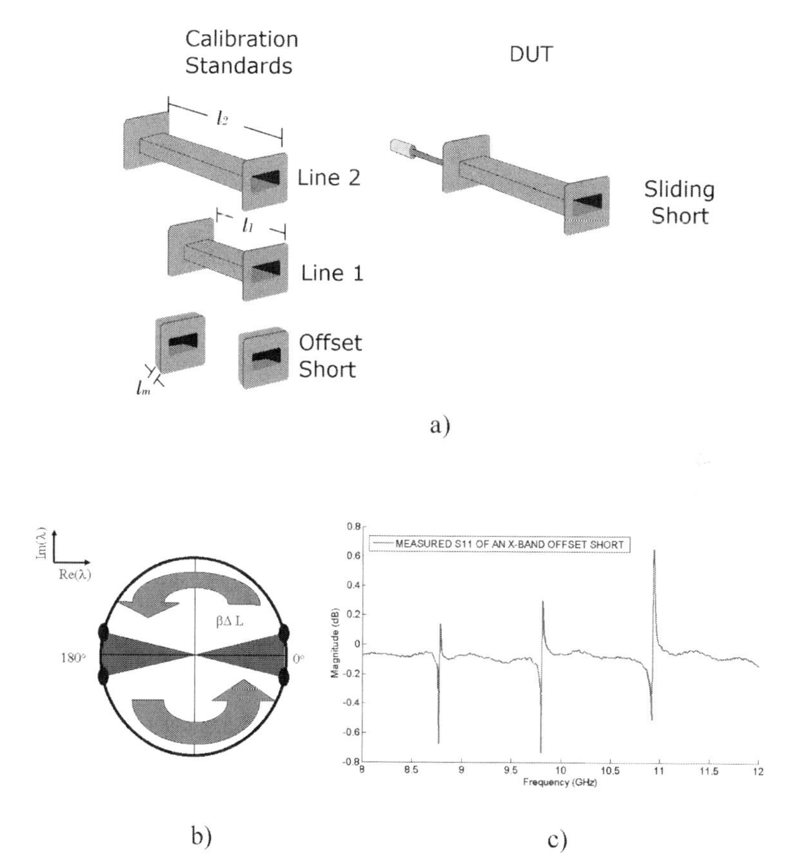

Figure 4.7 (a) Rectangular waveguide TRL calibration structures, (b) illustration of conditions at which non-singularities in TRL occur, and (c) example of calibration results showing spikes due to crossings by $k \cdot 180^\circ$ in angle of λ.

4.5 Generalized Theory of the TRM Calibration Technique

The thru-reflect-match (TRM) calibration technique is achieved by replacing the line used in the TRL by a broadband load. The TRM technique uses as calibration structures the through connection of the VNA ports (thru), a symmetrical reflecting load (reflect), and a pair of loads of impedance close to the measuring system impedance (match), as shown in Figure 4.8.

The TRM calibration technique was introduced in [22, 23] using the T-parameters formalism. In [22, 23], the loads used as match standard were assumed as symmetrical loads of impedance identical to Z_0, thus referring the calibration to the impedance of the load used as calibration element. When

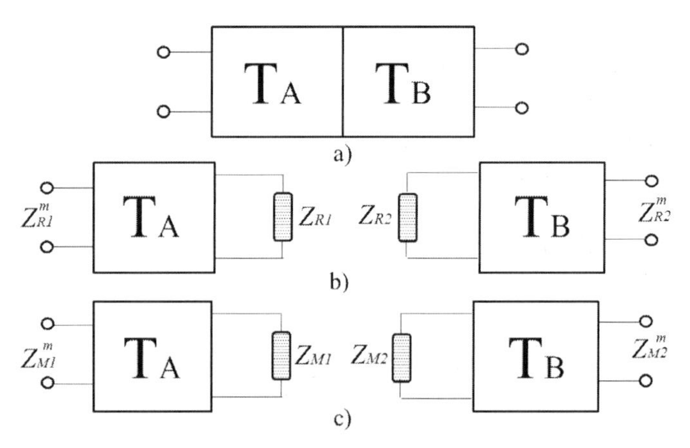

Figure 4.8 TRM calibration structures: (a) thru, (b) reflecting load, and (c) symmetrical/non-symmetrical load of arbitrary impedance. Ref [65]. © 2015 IEEE. Reprinted with permission.

dealing with planar structures (e.g., microstrip or coplanar waveguide), the value of the load impedance may be frequency-dependent [94]. Furthermore, as the frequency increases, the symmetry condition of the loads used as match standard may become difficult to preserve, thus affecting the reference impedance definition and reducing the calibration accuracy [43].

A TRM calibration procedure allowing the use of symmetrical loads of frequency-dependent impedance as the match standard was reported in [94]. The TRM of [94] uses the theory reported in [51] to transform the calibration reference impedance from a frequency-dependent impedance Z_M to the measurement system impedance Z_0 in a step after the calibration.

Experimental results regarding the implementation of a TRM-like calibration technique (LRM+), in which the use of non-symmetrical loads as the match standard is allowed, were reported in [24, 19]. Moreover, a procedure for calibrating the VNA, using a known transmitting calibration standard (thru), a symmetrical unknown double one-port standard (reflect), and a double (symmetrical or non-symmetrical) one-port standard (match) was patented in [38]. Nevertheless, in the procedure of [38], the value of the reflection coefficient of the match standard must be different from zero. Therefore, it does not allow the use of loads of impedance identical to Z_0.

The TRM procedure described in this section [65] uses the ABCD-parameters matrix formalism to develop a generalized TRM calibration theory, allowing the use of either symmetrical or non-symmetrical loads of arbitrary impedance as match standard. This procedure determines the $\overline{\frac{A_X}{C_X}}$

and $\overline{B_X}$ terms by combining the measurements of the thru with the measurement of a broadband load of arbitrary impedance. In this TRM procedure, the non-unity elements of the matrix $\mathbf{T_Z}$ (Z_A and Z_B) are identical to the impedance of the loads used as *match* standards in ports 1 and 2 of the VNA, respectively ($Z_A = Z_{M1}$ and $Z_B = Z_{M2}$), which may be either identical or different and of arbitrary value.

According to the analysis presented in Section 4.3.2, the impedance at the input of port P ($P = 1, 2$) of the VNA Z^m_{MP}, when it is loaded with a load of impedance Z_{MP}, may be expressed as

$$Z^m_{M1} = \frac{\overline{A_X}\,(Z_{M1} - Z_A) - \overline{B_X}\,(Z_{M1} + Z_B)}{\overline{C_X}\,(Z_{M1} - Z_A) - (Z_{M1} + Z_B)}, \tag{4.43}$$

$$Z^m_{M2} = \frac{\overline{B_Y}\,(Z_{M2} - Z_B) + (Z_{M2} + Z_A)}{\overline{A_Y}\,(Z_{M2} - Z_B) + \overline{C_Y}\,(Z_{M2} + Z_A)}. \tag{4.44}$$

Thus, by substituting $Z_A = Z_{M1}$ and $Z_B = Z_{M2}$ in eqn (4.43) and (4.44), one has

$$\overline{B_X} = Z^m_{M1}, \tag{4.45}$$

$$\overline{C_Y} = \frac{1}{Z^m_{M2}}. \tag{4.46}$$

The value of $\frac{\overline{A_X}}{\overline{C_X}}$ may be determined by substituting eqn (4.13) in eqn (4.46) as

$$\frac{\overline{A_X}}{\overline{C_X}} = \frac{p_{12} - p_{11} Z^m_{M2}}{p_{22} - p_{21} Z^m_{M2}}. \tag{4.47}$$

In the TRM technique, the $\overline{C_X}$ term is calculated as presented in Section 4.3.2 by using the measurement of a pair of reflecting loads.

The usefulness of the described TRM method is shown in Figures 4.9 and 4.10, in which a pair of loads of different impedance are used to calibrate a two-port VNA using the TRM technique. In this example, the impedance of the load used as *match* standard at port 1 is close to 50 Ω with inductive imaginary part. The load used at port 2, on the other hand, is a shifted load that is made by adding a piece of transmission line between the lumped load and the calibration plane, thus causing the impedance of the load to be notably different from the load used at port 1. These loads, along with the thru and the reflecting loads (short), are included in the CM05 calibration substrate from *J Micro Technology*.

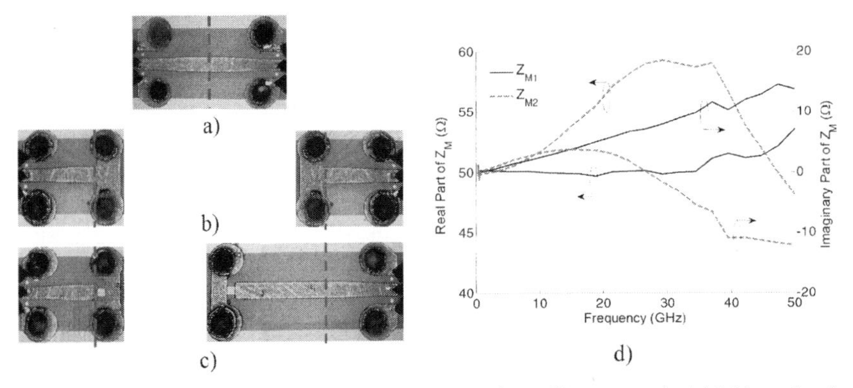

Figure 4.9 CM05 TRM calibration structures: (a) thru, (b) symmetrical highly reflecting loads, (c) non-symmetrical broadband loads, and (d) real and imaginary parts of the impedance of the loads used as match standards in ports 1 and 2 of the VNA. Red dotted lines denote the location of the calibration reference plane [65]. Ref [65]. © 2015 IEEE. Reprinted with permission.

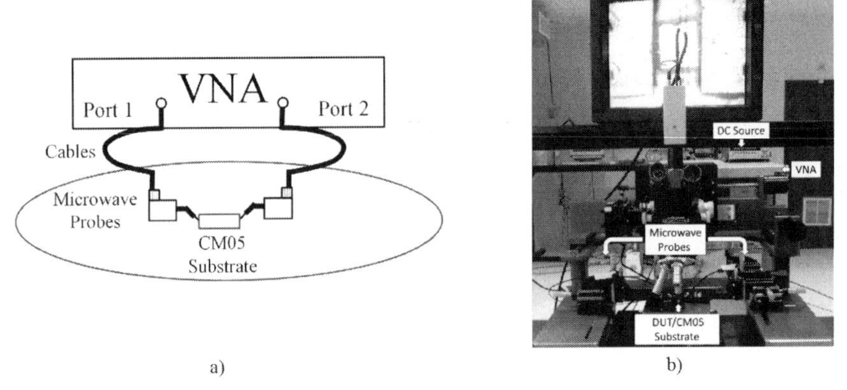

Figure 4.10 (a) Measurement description and (b) actual test system used to measure an HFET transistor after TRM (and TRL) calibration [65]. Ref [65]. © 2015 IEEE. Reprinted with permission.

As a DUT, an HFET transistor was measured after calibration is completed and the accuracy of the implemented TRM calibration was assessed by comparing the measured S-parameters against the S-parameters corrected by using the reference multi-line TRL calibration. The TRL calibration was referred to the measuring system impedance using the value of the line's characteristic impedance [72]. The characteristic impedance of the transmission line used in the TRL calibration was determined from its propagation constant and capacitance [52, 93].

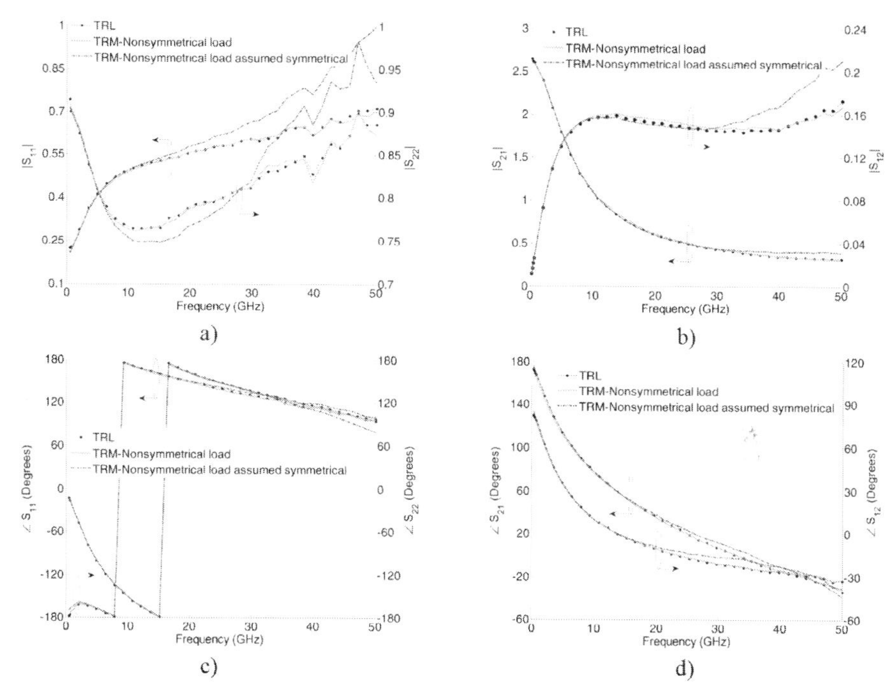

Figure 4.11 Measured S-parameters of an HFET transistor calibrated using TRM and TRL calibrations [65]. Ref [65]. © 2015 IEEE. Reprinted with permission.

Figure 4.11 shows the S-parameters of the DUT corrected using the *generalized theory of the TRM calibration* and the TRL calibration. The high correlation observed between the S-parameters corrected using the TRL calibration and the TRM validates this theory. The S-parameters corrected using the TRM calibration assuming the asymmetrical loads used as match standards as symmetrical (i.e., by assuming that the impedance of the load connected at port 2 is identical to the impedance of the load connected at port 1 of the VNA) are also shown. The obtained results show that the accuracy of the TRM is degraded when the existing asymmetry of the loads used as match standard is not accounted for in the calibration procedure.

4.6 The TRRM Calibration Technique

The thru-reflect-reflect-match (TRRM) calibration technique uses as calibration elements the through connection of the VNA ports (thru), a pair of symmetrical loads of very low impedance (short circuits), a pair of symmetrical loads of very high impedance (open circuits), and a one-port load of impedance close to Z_0 (match). Figure 4.12 depicts the set of calibration structures used in the TRRM calibration.

Unlike the TRM technique, in which it is required to know the impedance of two loads over the frequency band of interest, TRRM only uses one load of known impedance connected at one of the VNA ports, thus avoiding problems due to asymmetry between the loads used as match standards in the TRM procedure.

The TRRM technique was reported for the first time in [16]; nevertheless, the first TRRM mathematical formulation was presented in [68]. The TRRM procedure reported in [68] was developed using T-parameters; it allowed referring the calibration to either the impedance of the load used as match standard or to Z_0. Then, in [42] a TRRM procedure using ABCD-parameters was developed (ABCD-parameters are not defined as a function of a reference impedance).

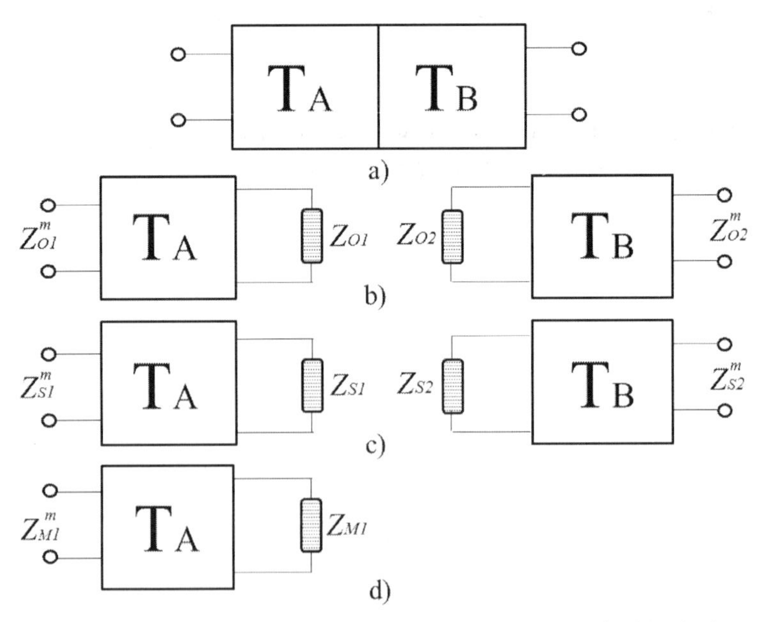

Figure 4.12 TRRM calibration structures: (a) thru, (b) a symmetrical load of very high impedance, (c) a symmetrical load of very low impedance, and (d) a one-port load of known impedance.

As in the TRM procedure, the TRRM uses the measurement of a load of known impedance Z_{M1} connected at port 1 of the VNA for determining the $\overline{B_X}$ term. Eqn (4.43) denotes the impedance at the input of port 1 of the VNA, Z_{M1}^m, when it is loaded with a load of impedance Z_{M1}

Unlike the TRM technique, in which the non-unity elements of the matrix $\mathbf{T_Z}$ (Z_A and Z_B) correspond to the impedance of the loads connected at ports 1 and 2 of the VNA, in the TRRM these terms are chosen to be identical to the impedance of the load connected at port 1. By substituting $Z_A = Z_{M1}$ in eqn (4.43), the following expression is obtained:

$$\overline{B_X} = Z_{M1}. \tag{4.48}$$

Since in the TRRM only one broadband load is used, $\frac{\overline{A_X}}{\overline{C_X}}$ cannot be determined by using the measurement of a load connected at port 2 as in the TRM. Instead, $\frac{\overline{A_X}}{\overline{C_X}}$ is calculated by using the thru and two pairs of reflecting loads, as presented next.

According to the theory presented in Section 4.2, by using the measurement of a symmetrical reflecting load, the following expression may be derived:

$$\overline{C_X}^2 = \left(\overline{A_Y} \cdot \overline{C_X}\right) \eta_{R1}/\eta_{R2}, \tag{4.49}$$

where *R* = *O*, *S* represents either the use of a load of very high impedance or a load of very low impedance. Then, by combining eqn (4.11) with eqn (4.49), the following expression may be derived:

$$\overline{C_X}^2 = \frac{h_R}{\left(\frac{\overline{A_X}}{\overline{C_X}}\right)^2 \cdot a_{2,R} + \left(\frac{\overline{A_X}}{\overline{C_X}}\right) \cdot a_{1,R} + a_{0,R}}, \tag{4.50}$$

where

$$\begin{aligned}
h_R &= \left(Z_{R1}^m - \overline{A_X}\right)\left(Z_{R2}^m - \overline{B_Y}/\overline{A_Y}\right)\left(p_{11} - \overline{B_X} p_{21}\right), \\
a_{0,R} &= Z_{R1}^m \left(Z_{R2}^m p_{11} - p_{12}\right), \\
a_{1,R} &= Z_{R1}^m \left(Z_{R2}^m p_{21} - p_{22}\right) + \left(Z_{R2}^m p_{11} - p_{12}\right), \\
a_{2,R} &= Z_{R2}^m p_{21} - p_{22}.
\end{aligned}$$

Since TRRM uses two types of reflecting loads, it is possible to generate two expressions similar to eqn (4.50). In the case of a load of very high impedance, one has

$$\overline{C_X}^2 = \frac{h_O}{\left(\frac{\overline{A_X}}{\overline{C_X}}\right)^2 \cdot a_{2,O} + \left(\frac{\overline{A_X}}{\overline{C_X}}\right) \cdot a_{1,O} + a_{0,O}}. \tag{4.51}$$

Meanwhile, for a load of very low impedance, one obtains the following expression:

$$\overline{C_X}^2 = \frac{h_S}{\left(\frac{\overline{A_X}}{\overline{C_X}}\right)^2 \cdot a_{2,S} + \left(\frac{\overline{A_X}}{\overline{C_X}}\right) \cdot a_{1,S} + a_{0,S}}. \tag{4.52}$$

To be consistent, the value of $\overline{C_X}$ in eqn (4.51) has to be identical to the value of $\overline{C_X}$ in eqn (4.52). Thus, by equating eqn (4.51) to eqn (4.52) and after some algebra, the following quadratic equation may be obtained:

$$\left(\frac{\overline{A_X}}{\overline{C_X}}\right)^2 \left(a_{2,S}\frac{h_O}{h_S} - a_{2,O}\right) + \left(\frac{\overline{A_X}}{\overline{C_X}}\right)\left(a_{1,S}\frac{h_O}{h_S} - a_{1,O}\right) + \left(a_{0,S}\frac{h_O}{h_S} - a_{0,O}\right) = 0. \tag{4.53}$$

The value of $\frac{\overline{A_X}}{\overline{C_X}}$ can then be determined from eqn (4.53) as

$$\frac{\overline{A_X}}{\overline{C_X}} = \frac{-\kappa_1 \pm \sqrt{\kappa_1^2 - 4\kappa_0\kappa_2}}{2\kappa_2}, \tag{4.54}$$

where $\kappa_i = a_{i,S}\left(\frac{h_O}{h_S}\right) - a_{i,O}$; $i = 0, 1, 2$.

As in the TRL and TRM, the value of $\overline{C_X}$ may be determined using the measurement of a pair of symmetrical reflecting loads, either open circuits or short circuits.

Figure 4.13 shows an example of the usefulness of the TRRM calibration method. The S-parameters of an HFET transistor corrected using the TRRM procedure are compared to the S-parameters corrected using the reference multi-line TRL technique. The load included in the calibration substrate used for such comparison (CM05 from Jc Micro) is highly inductive (close to 44 pH). The high correlation observed between both sets of S-parameters validates the accuracy of the TRRM method even if highly inductive loads are used as standard.

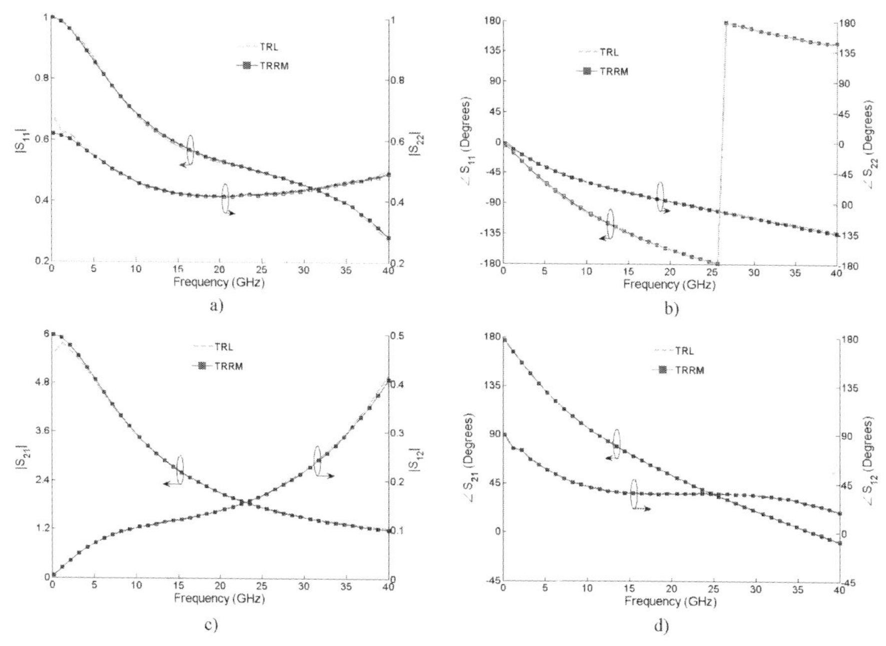

Figure 4.13 Measured S-parameters of an HFET transistor calibrated using TRRM and TRL calibrations.

4.7 The LZZ Calibration Technique

The line, offset-open, offset-short (LZZ) calibration technique [31, 63, 64, 62], presented in this section, uses as standards: a known transmission line, a pair of unknown offset-open circuits, and a pair of unknown offset-short circuits (Figure 4.14). Unlike TRM and TRRM, in the LZZ calibration, the use of a precisely characterized load is not required. Moreover, since the LZZ calibration does not use multiple transmission lines for calculating the calibration terms, it may be implemented using fixed spacing structures.

According to the theory presented in Section 4.2, the value of the non-unity elements of the $\mathbf{T_Z}$ matrix (Z_A and Z_B) are identical to the impedance of one of the structures used as calibration elements in a determined calibration. The unique requirement is that the values of Z_A and Z_B allow the matrix $\mathbf{T_Z}$ to be non-singular. In this order, Z_A and Z_B cannot be either the impedance of a short circuit (ideally zero) or the impedance of an open circuit (ideally infinity). In the LZZ, Z_A and Z_B are then chosen to be identical to the characteristic impedance of the transmission line ($Z_A = Z_B = Z_L$).

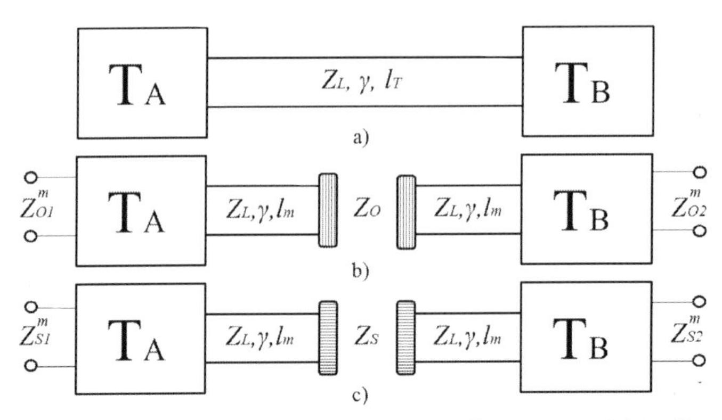

Figure 4.14 LZZ calibration structures: (a) line, (b) offset-open, and (c) offset-short. Ref [62]. © 2013 IEEE. Reprinted with permission.

The ABCD-parameters representation of the line structure shown in Figure 4.14(a) may be expressed as

$$\mathbf{M_L} = \mathbf{T_X T_\lambda T_Y}, \tag{4.55}$$

with the matrices $\mathbf{T_X}$ and $\mathbf{T_Y}$ previously defined in eqn (4.8) and (4.9).

Four of the seven terms of the matrices $\mathbf{T_X}$ and $\mathbf{T_Y}$ may be determined from eqn (4.11)–(4.14) using the measurement of the transmission line, as long as $\overline{\frac{A_X}{C_X}}$, $\overline{B_X}$, and $\overline{C_X}$ are known. The procedure for determining $\overline{\frac{A_X}{C_X}}$, $\overline{B_X}$, and $\overline{C_X}$ by combining the measurements of a line and two offset reflecting loads is presented next. The LZZ technique considers the loads used as calibration elements as offset loads (Figure 4.15).

In this analysis, an offset load connected at port $P = 1, 2$ is represented by a transmission line of length l_m terminated in a load of impedance Z_{RP}, with $R = O, S$ (open circuit or short circuit)[4]. The equivalent ABCD-parameters matrix of an offset load connected at port 1 of the VNA, $\mathbf{M_{R1}}$, may be expressed as

$$\mathbf{M_{R1}} = \mathbf{T_A T_{L_m} T_{R1}}. \tag{4.56}$$

Meanwhile, the equivalent ABCD-parameters matrix of an offset load connected at port 2 of the VNA, $\mathbf{M_{R2}}$, may be expressed as

$$\mathbf{M_{R2}} = \widetilde{\mathbf{T_B}}\mathbf{T_{L_m} T_{R2}}, \tag{4.57}$$

[4]The characteristic impedance and propagation constant of this line must be identical to the impedance of the line standard.

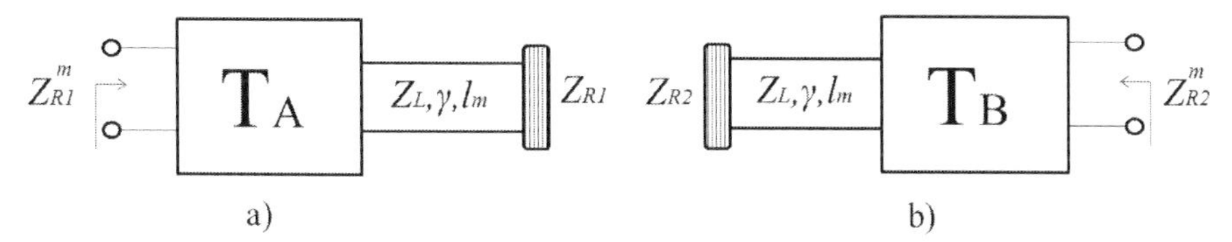

Figure 4.15 Equivalent structure of the offset loads connected at: (a) port 1 and (b) port 2 of the VNA. Ref [62]. © 2013 IEEE. Reprinted with permission.

with $\mathbf{T_{RP}}$ previously defined in eqn (4.18). Then, by substituting eqn (4.23) in eqn (4.57), $\mathbf{M_{R2}}$ may be expressed as follows:

$$\mathbf{M_{R2}} = \mathbf{X T_B^{-1} X T_{L_m} T_{R2}}. \tag{4.58}$$

In eqn (4.56)–(4.58), the matrix X is defined as in eqn 4.24, and $\mathbf{T_{L_m}}$ represents the ABCD-parameters matrix of a uniform transmission line of length l_m, which after diagonalization may be expressed as

$$\mathbf{T_{L_m}} = \mathbf{T_Z T_{\lambda_m} T_Z^{-1}}, \tag{4.59}$$

with $\mathbf{T_Z}$ defined in Section 4.2 and $\mathbf{T_{\lambda_m}}$ defined as

$$\mathbf{T_{\lambda_m}} = \begin{bmatrix} \lambda_{L_m} & 0 \\ 0 & \lambda_{L_m}^{-1} \end{bmatrix}, \tag{4.60}$$

where $\lambda_m = e^{-\gamma l_m}$. By solving eqn (4.8) and (4.9) for $\mathbf{T_A}$ and $\mathbf{T_B}$, respectively, and substituting the resulting expressions in eqn (4.56) and (4.58), one has

$$\mathbf{M_{K1}} = \mathbf{T_X T_{\lambda_m} T_Z T_{R1}}, \tag{4.61}$$

$$\mathbf{M_{K2}} = \mathbf{X T_Y^{-1} T_Z^{-1} X T_{L_m} T_{R2}}. \tag{4.62}$$

Then, by substituting eqn (4.59) in eqn (4.61) and (4.62) and after some algebra, the impedance at the input of port $P = 1, 2$, Z^m_{RP}, when it is loaded with an offset load of impedance Z_{RP}, may be expressed as

$$Z^m_{R1} = \frac{\overline{A_X}\left(Z_{R1} - Z_L\right)\lambda_m - \overline{B_X}\left(Z_{R1} + Z_L\right)\lambda_m^{-1}}{\overline{C_X}\left(Z_{R1} - Z_L\right)\lambda_m - \left(Z_{R1} + Z_L\right)\lambda_m^{-1}}, \tag{4.63}$$

$$Z^m_{R2} = \frac{\overline{B_Y}\left(Z_{R2} - Z_L\right)\lambda_m + \left(Z_{R2} + Z_L\right)\lambda_m^{-1}}{\overline{A_Y}\left(Z_{R2} - Z_L\right)\lambda_m + \overline{C_Y}\left(Z_{R2} + Z_L\right)\lambda_m^{-1}}. \tag{4.64}$$

The $\frac{\overline{A_X}}{\overline{C_X}}$ and $\overline{B_X}$ terms are determined by comparing the value of Z_L to the value of Z_{RP}. Two cases may be distinguished:

1. The following expressions may be obtained from eqn (4.63) and (4.64) when a load of impedance Z_{RP}, such that $|Z_{RP}| \gg |Z_L|$, $R = O$ (open circuit), is connected at the VNA ports

$$\overline{C_X} = \frac{1}{\lambda_m^2} \frac{Z_{O1}^m - \overline{B_X}}{Z_{O1}^m - \frac{\overline{A_X}}{\overline{C_X}}}, \tag{4.65}$$

$$\overline{A_Y} = \frac{1}{\lambda_m^2} \frac{Z_{O2}^m \overline{C_Y} - 1}{\frac{\overline{B_Y}}{\overline{A_Y}} - Z_{O2}^m}. \tag{4.66}$$

2. Similarly, the following expressions may be derived from eqn (4.63) and (4.64) when a load of impedance Z_{RP}, such that $|Z_{RP}| \ll |Z_L|$, $R = S$ (short circuit), is connected at the VNA ports

$$\overline{C_X} = -\frac{1}{\lambda_m^2} \frac{Z_{S1}^m - \overline{B_X}}{Z_{S1}^m - \frac{\overline{A_X}}{\overline{C_X}}}, \tag{4.67}$$

$$\overline{A_Y} = -\frac{1}{\lambda_m^2} \frac{Z_{S2}^m \overline{C_Y} - 1}{\frac{\overline{B_Y}}{\overline{A_Y}} - Z_{S2}^m}. \tag{4.68}$$

By equating eqn (4.65) and (4.67), the following expression may be derived:

$$\eta_1 a_1 + \eta_2 a_2 = W_a, \tag{4.69}$$

where

$$\eta_1 = \frac{\overline{A_X}}{\overline{C_X}} + \overline{B_X}, \tag{4.70}$$

$$\eta_2 = \frac{\overline{A_X}}{\overline{C_X}} \cdot \overline{B_X}, \tag{4.71}$$

$$a_1 = Z_{C1}, \tag{4.72}$$

$$a_2 = -2, \tag{4.73}$$

$$W_a = 2Z_{D1}. \tag{4.74}$$

A second expression, similar to eqn (4.69), may be derived by equating eqn (4.66) with eqn (4.68) and using eqn (4.11)–(4.13) as

$$\eta_1 b_1 + \eta_2 b_2 = W_b, \tag{4.75}$$

where

$$b_1 = 2(Z_{D2}p_{11}p_{22} + p_{12}p_{22}) - Z_{C2}(p_{11}p_{22} + p_{12}p_{21}), \quad (4.76)$$
$$b_2 = -2(Z_{D2}p_{21}^2 + p_{22}^2 - Z_{C2}p_{21}p_{22}), \quad (4.77)$$
$$W_b = 2(Z_{D2}p_{11}^2 + p_{12}^2 - Z_{C2}p_{11}p_{12}), \quad (4.78)$$

with Z_{CP} and Z_{DP} defined as follows:

$$Z_{CP} = Z_{OP}^m + Z_{SP}^m, \quad (4.79)$$
$$Z_{DP} = Z_{OP}^m \cdot Z_{SP}^m. \quad (4.80)$$

The expressions shown in eqn (4.69) and (4.75) form a system of linear simultaneous equations with two unknowns, η_1 and η_2. Solving this set of equations, one has

$$\eta_1 = \frac{W_a b_2 - W_b a_2}{a_1 b_2 - a_2 b_1}, \quad (4.81)$$

$$\eta_2 = \frac{W_b a_1 - W_a b_1}{a_1 b_2 - a_2 b_1}. \quad (4.82)$$

Then, by using the definitions for η_1 and η_2 given in eqn (4.70) and (4.71), the following quadratic equations with roots $\overline{\frac{A_X}{C_X}}$ and $\overline{B_X}$ result

$$\left(\overline{\frac{A_X}{C_X}}\right)^2 + \overline{\frac{A_X}{C_X}}\eta_1 + \eta_2 = 0, \quad (4.83)$$

$$\left(\overline{B_X}\right)^2 + \overline{B_X}\eta_1 + \eta_2 = 0, \quad (4.84)$$

with solutions given by

$$\overline{\frac{A_X}{C_X}}, \overline{B_X} = \frac{\eta_1 \pm \sqrt{\eta_1^2 - 4\eta_2}}{2}. \quad (4.85)$$

As explained in Section 4.4, it is mandatory that $\overline{\frac{A_X}{C_X}} \neq \overline{B_X}$. Therefore, one of the roots of eqn (4.85) corresponds to $\overline{\frac{A_X}{C_X}}$ and the other root corresponds to $\overline{B_X}$. The appropriate root for $\overline{\frac{A_X}{C_X}}$ and $\overline{B_X}$ may be chosen by using the following criterion:

$$\mathrm{Re}\left\{\overline{B_X}\right\} > \mathrm{Re}\left\{\overline{\frac{A_X}{C_X}}\right\}. \quad (4.86)$$

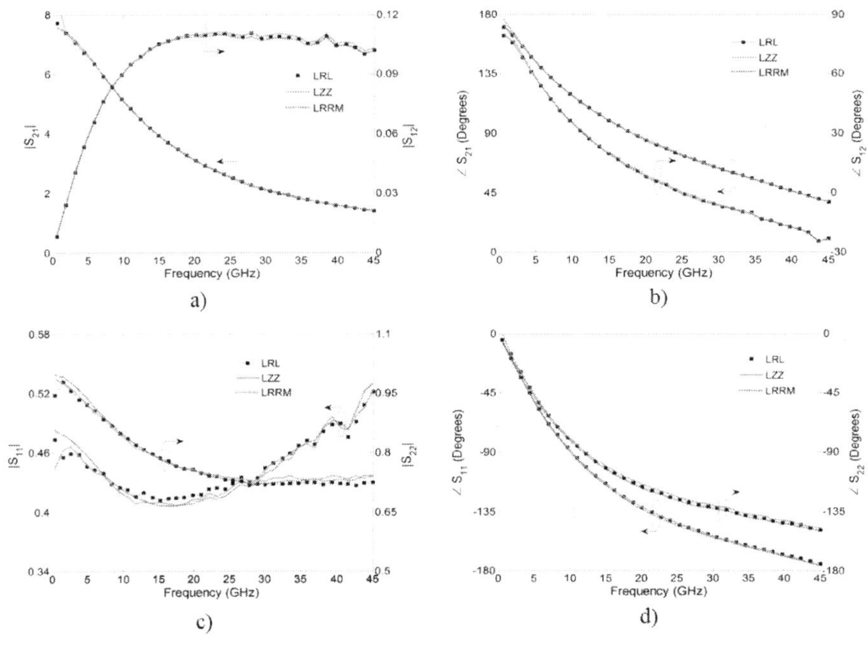

Figure 4.16 S-parameters of an HFET transistor calibrated using the LZZ, LRRM, and TRL calibrations. Ref [62]. © 2013 IEEE. Reprinted with permission.

Regarding the $\overline{C_X}$ term, it is calculated as in the TRL, TRM, and TRRM procedures, by using the measurement of a symmetrical reflecting load.

$$\overline{C_X} = \pm\sqrt{\left(\overline{A_Y} \cdot \overline{C_X}\right) \eta_{R1}/\eta_{R2}}. \tag{4.87}$$

Figure 4.16 shows a comparison of the S-parameters of a transistor, measured on-wafer, corrected using the LZZ, TRL, and LRRM techniques. The results show that the LZZ calibration may provide accuracy similar to the one provided by the enhanced LRRM technique [42], even though the LZZ uses less calibration structures. The TRL technique is used as a common reference for both LZZ and LRRM techniques.

4.8 The LZZM Calibration Technique

The theory of the LZZ technique was extended in [63] to develop a calibration procedure, which is comparable to the LRRM technique: the line, open, short, unknown load (LZZM). The LZZM uses as standards a known transmission line, two pairs of unknown reflecting loads, and a load of unknown impedance (match) connected at port 1 of the VNA, as shown in Figure 4.17.

In the LZZ technique, the calculation of the terms $\overline{\frac{A_X}{C_X}}$ and $\overline{B_X}$ implies appropriately choosing the sign of a root of a quadratic equation with a root selection criterion that may fail at high frequencies [63]. The use of a load of impedance value close to the impedance of the transmission line used as standard in the LZZM allows developing an analytic procedure to determine the appropriate root.

In eqn (4.88), an expression for the impedance at the input of port 1 of the VNA Z_{M1}^{m} when it is loaded with an offset load is presented

$$Z_{M1}^{m} = \frac{\overline{A_X}\left(Z_M - Z_L\right)\lambda_m - \overline{B_X}\left(Z_M + Z_L\right)\lambda_m^{-1}}{\overline{C_X}\left(Z_M - Z_L\right)\lambda_m - \left(Z_M + Z_L\right)\lambda_m^{-1}}. \tag{4.88}$$

Whether the value of Z_M is close to Z_L, it is possible to note that Z_{M1}^{m} is close to the value of $\overline{B_X}$; thus, the following criterion may be derived:

$$\left|Z_{M1}^{m} - \overline{B_X}\right| < \left|Z_{M1}^{m} - \overline{A_X}/\overline{C_X}\right|. \tag{4.89}$$

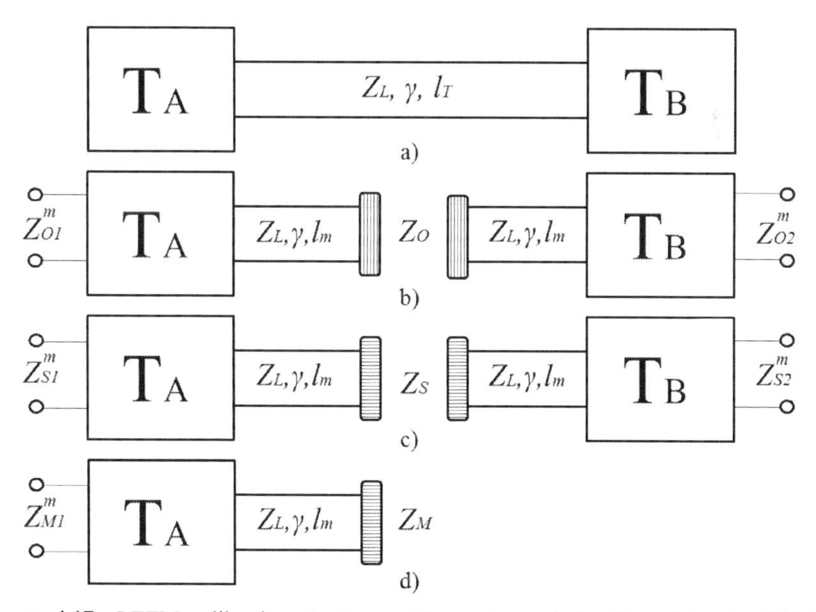

Figure 4.17 LZZM calibration structures: (a) one transmission line, (b) pair of highly reflecting loads (open), (c) pair of highly reflecting loads (short), and (d) one broadband load (match). Ref [63]. © 2014 IEEE. Reprinted with permission.

The TRRM/LRRM calibration procedure is one of the most commonly used techniques for correcting VNA systematic errors. In order to accurately calibrate the VNA by using the TRRM/LRRM technique, the impedance of a load (Z_M) connected at one of the ports of the VNA, *match*, has to be known prior to the calibration. The LZZM differs from TRRM/LRRM methods in the sense that it allows using loads of unknown frequency-dependent impedance.

As in the LZZ procedure, in the LZZM, the $\overline{\frac{A_X}{C_X}}$ and $\overline{B_X}$ terms are calculated from eqn (4.81)–(4.82) and eqn (4.85), by combining the measurements of the transmission line and the offset reflecting loads. Then, the $\overline{C_X}$ term is calculated by using the measurement of a load of impedance Z_M connected at port 1 of the VNA, as follows.

From eqn (4.88), the following expression may be obtained:

$$\overline{C_X} = \frac{1}{\lambda_m} \frac{Z_M/Z_L - 1}{Z_M/Z_L + 1} \cdot \frac{Z_{M1}^m - \overline{B_X}}{Z_{M1}^m - \overline{\frac{A_X}{C_X}}}. \tag{4.90}$$

It is possible to observe that for determining the value of $\overline{C_X}$, it is necessary to know the ratio Z_M/Z_L. This ratio may be determined by combining the measurement of a load with the measurement of an open circuit as presented next. By equating eqn (4.90) with eqn (4.65), one has

$$\frac{Z_M}{Z_L} = \frac{1+\xi}{1-\xi}, \tag{4.91}$$

$$\xi = \frac{\left(Z_{O1}^m - \overline{B_X}\right)\left(Z_{M1}^m \overline{\frac{A_X}{C_X}}\right)}{\left(Z_{O1}^m - \overline{B_X}\right)\left(Z_{M1}^m \overline{\frac{A_X}{C_X}}\right)}. \tag{4.92}$$

Note that Z_M has to be different from Z_L; otherwise, the use of the load provides redundant information (i.e., $Z_{M1}^m = \overline{B_X}$) and the value of $\overline{C_X}$ cannot be determined from eqn (4.90). In such a case, $\overline{C_X}$ has to be determined from the measurement of a symmetrical reflecting load.

4.9 The OSMT Calibration Technique

The open-short-match-thru (OSMT) calibration is based on the three terms error model and uses the measurement of three loads of different (known) impedance at one port of the VNA. In this section, the OSMT technique

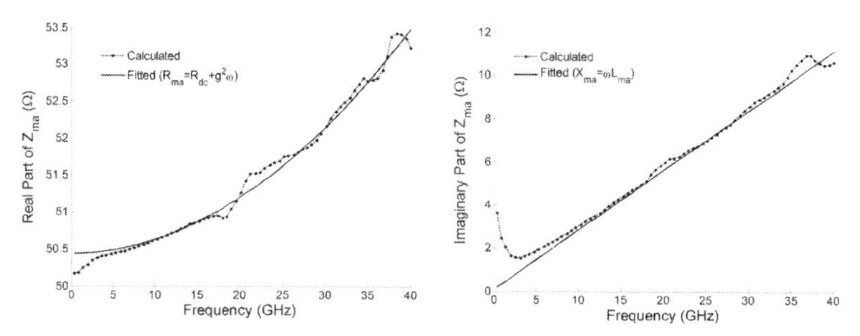

Figure 4.18 LZZM load. Ref [63]. © 2014 IEEE. Reprinted with permission.

is developed using the same ABCD-parameters matrix formalism used in previous sections. The case in which the loads are measured at port 1 is shown first.

The loads used in OSMT calibration technique may be represented as offset loads, as shown in Figure 4.20. The offset loads connected at port $N = 1, 2$ are represented as a transmission line of impedance Z_L, propagation constant γ, and length l_{KN}, where $K = O, S, M$. Hence, the impedance at the input of port N when it is loaded with an offset load may be represented as

$$Z_{K1}^m = \frac{\overline{A_X}\left(Z_{K1} - Z_L\right)\lambda_{K1} - \overline{B_X}\left(Z_{K1} + Z_L\right)\lambda_{K1}^{-1}}{\overline{C_X}\left(Z_{K1} - Z_L\right)\lambda_{K1} - \left(Z_{K1} + Z_L\right)\lambda_{K1}^{-1}}, \tag{4.93}$$

and

$$Z_{K2}^m = \frac{\overline{B_Y}\left(Z_{K2} - Z_L\right)\lambda_{K2} + \left(Z_{K2} + Z_L\right)\lambda_{K2}^{-1}}{\overline{A_Y}\left(Z_{K2} - Z_L\right)\lambda_{K2} + \overline{C_Y}\left(Z_{K2} + Z_L\right)\lambda_{K2}^{-1}}, \tag{4.94}$$

where $\lambda_{KN} = e^{-\gamma l_{KN}}$ represent the phase shift of the offset load connected at port N. After some algebra, eqn (4.93) and (4.94) can be rearranged as

$$\overline{A_X}\left(G_{K1}\lambda_{K1}^2\right) - \overline{B_X} - \overline{C_X}\left(Z_{K1}^m G_{K1}\lambda_{K1}^2\right) = -Z_{K1}^m, \tag{4.95}$$

and

$$\overline{A_Y}\left(G_{K2}\lambda_{K2}^2\right) - \overline{B_Y}\left(Y_{K2}^m G_{K2}\lambda_{K2}^2\right) + \overline{C_Y} = Y_{K2}^m, \tag{4.96}$$

where $G_{KN} = \left(Z_{KN} - Z_L\right) / \left(Z_{KN} + Z_L\right)$. By using the measurement of a high-impedance load (open circuit, O), a low-impedance load (short circuit,

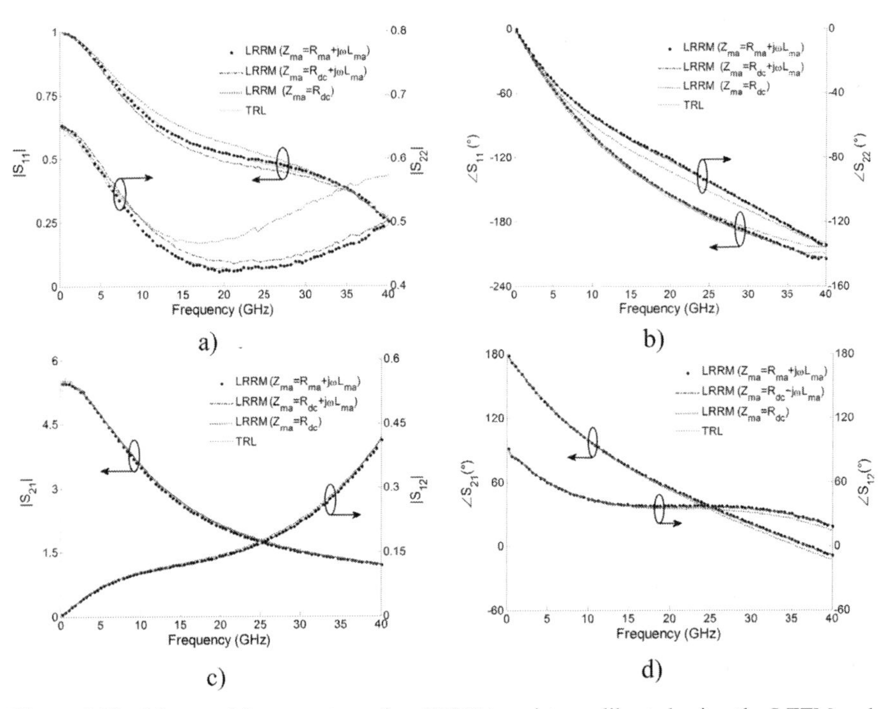

Figure 4.19 Measured S-parameters of an HFET transistor calibrated using the LZZM and LRRM calibrations. Ref [63]. © 2014 IEEE. Reprinted with permission.

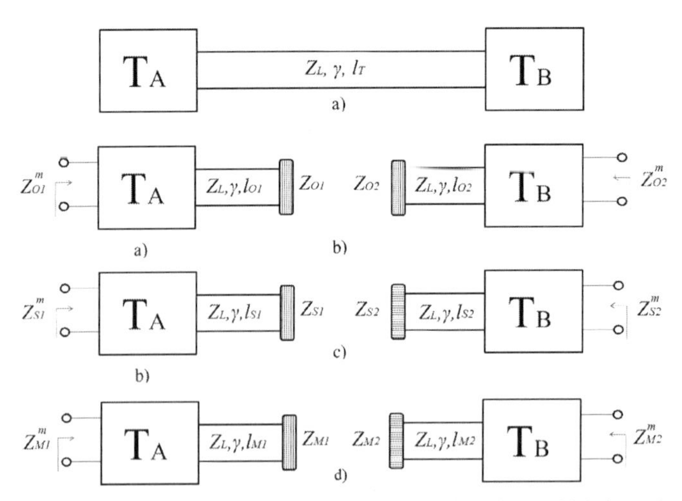

Figure 4.20 OSM calibration structures: (a) an offset load of very high impedance, (b) an offset load of very low impedance (short, *S*), and (c) an offset load of impedance close to the measuring system impedance (match, *M*).

S) and a load of impedance value close to the measuring system impedance (matched load, M), at port 1, the following matrix solution for $\overline{A_X}$, $\overline{B_X}$, and $\overline{C_X}$ may be derived

$$\begin{bmatrix} \overline{A_X} \\ \overline{B_X} \\ \overline{C_X} \end{bmatrix} = \begin{bmatrix} G_{O1}\lambda_{O1}^2 & -1 & -Z_{O1}^m G_{O1}\lambda_{O1}^2 \\ G_{S1}\lambda_{S1}^2 & -1 & -Z_{S1}^m G_{S1}\lambda_{S1}^2 \\ G_{M1}\lambda_{M1}^2 & -1 & -Z_{M1}^m G_{M1}\lambda_{M1}^2 \end{bmatrix}^{-1} \begin{bmatrix} -Z_{O1}^m \\ -Z_{S1}^m \\ -Z_{M1}^m \end{bmatrix}. \tag{4.97}$$

Similarly, by using the measurement of these three loads at port 2, the following matrix solution for $\overline{A_Y}$, $\overline{B_Y}$, and $\overline{C_Y}$ may be derived:

$$\begin{bmatrix} \overline{A_Y} \\ \overline{B_Y} \\ \overline{C_Y} \end{bmatrix} = \begin{bmatrix} G_{O2}\lambda_{O2}^2 & -Y_{O2}^m G_{O2}\lambda_{O2}^2 & 1 \\ G_{S2}\lambda_{S2}^2 & -Y_{S2}^m G_{S2}\lambda_{S2}^2 & 1 \\ G_{M2}\lambda_{M2}^2 & -Y_{M2}^m G_{M2}\lambda_{M2}^2 & 1 \end{bmatrix}^{-1} \begin{bmatrix} Y_{O2}^m \\ Y_{S2}^m \\ Y_{M2}^m \end{bmatrix}, \tag{4.98}$$

where $Y_{K2}^m = 1/Z_{K2}^m$. Finally, in order to complete the calibration, the value of the term $D_X D_Y$ must be determined. This term can be determined by using the two-port measurement of the transmission line depicted in Figure 4.20(d), as

$$D_X D_Y = \frac{\frac{\overline{A_X}}{\overline{C_X}} p_{22} - p_{12}}{\frac{\overline{A_X}}{\overline{C_X}} - \overline{B_X}} \lambda_L, \tag{4.99}$$

as described in Section 4.3.1. Note that in order to complete the calibration using this procedure, the propagation constant and length of the transmission line are required.

An alternative procedure to determine $D_X D_Y$ is based on the well-known SOLT with "unknown thru" calibration procedure [27]. By calculating the determinant of the matrix product describing the measurement of a transmission line,

$$\mathbf{M_L} = \mathbf{T_X} \mathbf{T_\lambda} \mathbf{T_Y}, \tag{4.100}$$

and considering that $\det(\mathbf{T_\lambda}) = 1$, the following equation to determine the term $D_X D_Y$ may be obtained:

$$D_X D_Y = \sqrt{\frac{det(\mathbf{M_L})}{(\overline{A_X} - \overline{B_X C_X})(\overline{A_Y} - \overline{B_Y C_Y})}}. \tag{4.101}$$

4.10 Sensitivity of Device Characterization Due to Uncertainty on Calibration Structures

As described in the previous section, in the TRL, TRM, and TRRM calibration techniques, there is a parameter that greatly impacts the accuracy of the calibration; such a parameter may be the impedance of a transmission line (TRL), the impedance of a one-port load (TRRM), or the impedance of a pair of loads (TRM). While there are other factors that may affect the accuracy of these calibrations, the mathematical description shown in previous sections demonstrates that in the final de-embedding (4.10), knowing the elements of the $\mathbf{T_Z}$ matrix is imperative for the accurate calculation of DUT's ABCD-parameters.

A common question within the microwave measurement and characterization community is related to the impact that errors in the knowledge of the impedance of a calibration structure may have on the accurate characterization of a DUT. In this section, the sensitivity of DUT characterization due to uncertainty of calibration structures is presented. While the mathematical formulation is general enough to cover different calibration procedures, for the sake of compactness, the examples shown are limited to the TRM calibration technique.

The TRL, TRM, and TRRM techniques use as calibration structures the through connection of the VNA ports, along with other calibration structures, that may include transmission lines, reflecting loads, and matched loads. The analysis presented here assumes that a zero-length thru is used as the through connection. It utilizes the ABCD-parameters matrix formalism for VNA calibration, presented in previous sections. Thus, as previously mentioned, the uncorrected ABCD-parameters matrix of a DUT, $\mathbf{M_D}$, may be expressed as a function of its actual ABCD-parameters, $\mathbf{T_D}$, as

$$\mathbf{M_D} = \mathbf{T_X}\mathbf{T_Z^{-1}}\mathbf{T_D}\mathbf{T_Z}\mathbf{T_Y}, \tag{4.102}$$

with $\mathbf{T_Z}$ defined as

$$\mathbf{T_Z} = \begin{bmatrix} -Z_C & Z_C \\ 1 & 1 \end{bmatrix}, \tag{4.103}$$

where $Z_C = Z_L$ (TRL), $Z_C = Z_M$ (TRM).

Similarly, as shown in Table 4.2, out of the eight terms included in $\mathbf{T_X}$ and $\mathbf{T_Y}$ matrices, seven terms may be determined by using uncalibrated measurements of a set of calibration structures. The TRM, TRL, and TRRM procedures do not require knowing the electrical behavior of

Table 4.2 Calculation of the calibration terms using the calibration structures of different calibration techniques. The corresponding values of the impedances Z_A and Z_B are also presented.

	$\overline{B_X}$	$\overline{\frac{A_X}{C_X}}$	$\overline{C_X}$	$\overline{A_Y}, \overline{B_Y}, \overline{C_Y}$ $D_X D_Y$
TRL $Z_A = Z_B = Z_L$	Thru-Line	Thru-Line	Open P1-Open P2/ Short P1-Short P2	Thru
TRM $Z_A = Z_{M1}$, $Z_B = Z_{M2}$	Load P1	Thru-Load P2	Open P1-Open P2/ Short P1-Short P2	Thru
TRRM $Z_A = Z_B = Z_{M1}$	Load P1	Thru, Open P1-Open P2, Short P1-Short P2	Open P1-Open P2/ Short P1-Short P2	Thru
LZZ $Z_A = Z_B = Z_L$	Line, Open P1-Short P1, Open P2-Short P2	Line, Open P1-Short P1, Open P2-Short P2	Open P1-Open P2/ Short P1-Short P2	Line
LZZM $Z_A = Z_B = Z_L$	Line, Open P1-Short P1, Open P2-Short P2	Line, Open P1-Short P1, Open P2-Short P2	Open P1-Open P2/ Short P1-Short P2	Line

the calibration structures to determine those seven calibration terms. This allows to isolate impact of Z_C through $\mathbf{T_Z}$, on the calculation of the DUT's ABCD-parameters.

Let us calculate the Z-parameters and Y-parameters of a DUT. By solving eqn 4.102 for $\mathbf{T_D}$, one has

$$\mathbf{T_D} = \mathbf{T_Z}\,\overline{\mathbf{M_D}}\,\mathbf{T_Z^{-1}} = \begin{bmatrix} a_{11} & Z_C\,a_{12} \\ Y_C\,a_{21} & a_{22} \end{bmatrix}, \tag{4.104}$$

where $Y_C = 1/Z_C$. The matrix $\overline{\mathbf{M_D}}$, which is determined solely by using measurements of a set of calibration structures, is defined as

$$\overline{\mathbf{M_D}} = \mathbf{T_X^{-1}}\,\mathbf{M_D}\,\mathbf{T_Y^{-1}} = \begin{bmatrix} m_{11} & m_{12} \\ m_{21} & m_{22} \end{bmatrix}, \tag{4.105}$$

with

$$a_{11} = \left[(m_{11} + m_{22}) - (m_{12} + m_{21})\right]/2, \tag{4.106}$$

$$a_{12} = \left[(m_{22} - m_{11}) + (m_{12} - m_{21})\right]/2, \tag{4.107}$$

$$a_{12} = \left[(m_{22} - m_{11}) - (m_{21} + m_{12})\right]/2, \tag{4.108}$$

$$a_{22} = \left[(m_{11} + m_{22}) + (m_{21} + m_{12})\right]/2. \tag{4.109}$$

Then, the corrected $Z-$ and Y-parameters matrices of a DUT may be determined by using two-port network parameters conversion procedures [30] as

$$\mathbf{Z_D} = \frac{Z_C}{a_{21}}\begin{bmatrix} a_{11} & \Delta_a \\ 1 & a_{22} \end{bmatrix}, \tag{4.110}$$

and

$$\mathbf{Y_D} = \frac{Y_C}{a_{12}}\begin{bmatrix} a_{22} & -\Delta_a \\ -1 & a_{11} \end{bmatrix}, \tag{4.111}$$

respectively, where $\Delta_a = a_{11}a_{22} - a_{12}a_{21}$.

The terms $a_{i,j}$, $i, j = 1, 2$, are determined solely from uncorrected measurements of a DUT and calibration structures. Thus, from eqn (4.110) and (4.111), it may be noted that the dependence of the corrected $Z-$ and Y-parameters of a DUT on the TRL, TRM, or TRRM calibration structures lies on the knowledge of Z_C.

An application of this analysis, published in [60], is the evaluation of the sensitivity of FET parasitic elements extraction due to uncertainty on

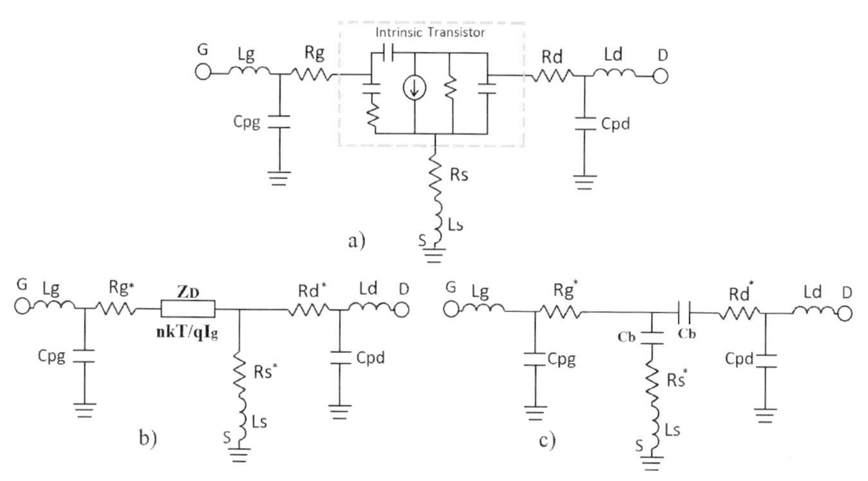

Figure 4.21 (a) Small-signal ECM of an FET; ECM (b) under floating drain at $V_{\mathrm{GS}} > V_{\mathrm{bi}} > 0$ (forward cold-FET) and (c) under $V_{\mathrm{DS}} = 0$ at $V_{\mathrm{GS}} < V_T$ (reverse cold-FET). Ref [60]. © 2021 IEEE. Reprinted with permission.

TRM calibration structures. Measurement-based equivalent circuit models (ECMs) are within the most common transistor modeling approaches. They are electrical representations of the FET's physical structure, in which measurements performed on the device to be modeled are used to extract the ECM parameters. There are several sources of uncertainty that contribute to the overall uncertainty in the modeling results, where one of the most common is the measurement uncertainty. Since ECM parameters are extracted from data measured using a calibrated VNA, it is expected that calibration errors impact the modeling results.

The FET small-signal ECM used in [60], comprising parasitic and intrinsic elements, is shown in Figure 4.21(a). The extraction of the parasitic elements starts with the calculation of the parasitic resistances and inductances. This is accomplished by using the Z-parameters of the circuit shown in Figure 4.21(b), which represents the FET behavior under floating drain and V_{GS} greater than the device's built-in voltage (V_{bi}). This bias condition is known as "forward cold-FET." Let $\mathbf{Z_F}$, defined as

$$\mathbf{Z_F} = \begin{bmatrix} Z_{11}^f & Z_{12}^f \\ Z_{21}^f & Z_{22}^f \end{bmatrix} = \frac{Z_C}{a_{21}^f} \begin{bmatrix} a_{11}^f & \Delta a_f \\ 1 & a_{22}^f \end{bmatrix}, \tag{4.112}$$

$\Delta a_f = a_{11}^f a_{22}^f - a_{12}^f a_{21}^f$, be the corrected Z-parameters matrix of the FET under forward cold-FET bias condition. At low frequencies, the real part of the Z-parameters of the circuit shown in Figure 4.21(b) may be expressed as

$$\mathrm{Re}\{Z_{11}^f\} = R_g^* + R_s^* + Z_D, \tag{4.113}$$

$$\mathrm{Re}\{Z_{12}^f\} + \mathrm{Re}\{Z_{21}^f\} = R_s^*, \tag{4.114}$$

$$\mathrm{Re}\{Z_{22}^f\} = R_d^* + R_s^*, \tag{4.115}$$

where the term $Z_D = nkT/qI_g$ represents the bias-dependent impedance of the Schottky barrier at the gate terminal, and the terms R_g^* , R_d^* , and R_s^* represent the FET parasitic resistances in series with the effect of the channel resistance on every device terminal. Similarly, at high frequencies, the imaginary part of the Z-parameters corresponding to this circuit are expressed as

$$Im\{Z_{11}^f\} = \omega(L_g + L_s), \tag{4.116}$$

$$Im\{Z_{12}^f\} = Im\{Z_{21}^f\} = \omega(L_s), \tag{4.117}$$

$$Im\{Z_{22}^f\} = \omega(L_d + L_s), \tag{4.118}$$

where ω represents the angular frequency. By neglecting the channel resistance, from eqn (4.113)–(4.115), the following expressions for the parasitic resistances are obtained:

$$Rg = \mathrm{Re}\{Z_M A_g\} = \mathrm{Re}(Z_C)\mathrm{Re}\{A_g\} - Im(Z_C)Im\{A_g\}, \tag{4.119}$$

$$Rs = \mathrm{Re}\{Z_M A_s\} = \mathrm{Re}(Z_C)\mathrm{Re}\{A_s\} - Im(Z_C)Im\{A_s\}, \tag{4.120}$$

$$Rd = \mathrm{Re}\{Z_M A_d\} = \mathrm{Re}(Z_C)\mathrm{Re}\{A_d\} - Im(Z_C)Im\{A_d\}, \tag{4.121}$$

where $A_g = (a_{11}^f - \Delta a_f)/a_{21}^f$, $A_s = \Delta a_f/a_{21}^f$, and $A_d = (a_{22}^f - \Delta a_f)/a_{21}^f$. Similarly, by using eqn (4.116)–(4.118), the following expressions for the parasitic inductances are obtained:

$$Lg = Im\{Z_M A_g\}/\omega = [\mathrm{Re}(Z_C)Im\{A_g\} - Im(Z_C)\mathrm{Re}\{A_g\}]/\omega, \tag{4.122}$$

$$Ls = Im\{Z_M A_s\}/\omega = [\mathrm{Re}(Z_C)Im\{A_s\} - Im(Z_C)\mathrm{Re}\{A_s\}]/\omega, \tag{4.123}$$

$$Ld = Im\{Z_M A_d\}/\omega = [\mathrm{Re}(Z_C)Im\{A_d\} - Im(Z_C)\mathrm{Re}\{A_d\}]/\omega. \tag{4.124}$$

The parasitic capacitances are extracted by biasing the FET at $V_{\mathrm{DS}} = 0V$ and V_{GS} lower than the device's threshold voltage (V_T). This bias condition is known as "reverse cold-FET". Under this condition, the FET behavior may be represented by the circuit shown in Figure 4.21(c). To extract the parasitic capacitances, the imaginary parts of the Y-parameters corresponding to this circuit at low frequencies are used. Thus, let Y_R, defined as

$$\mathbf{Y_R} = \begin{bmatrix} Y^r_{11} & Y^r_{12} \\ Y^r_{21} & Y^r_{22} \end{bmatrix} = \frac{Y_C}{a^r_{12}} \begin{bmatrix} a^r_{11} & \Delta a_r \\ 1 & a^r_{22} \end{bmatrix}, \tag{4.125}$$

with $\Delta a_r = a^r_{11}a^r_{22} \quad a^r_{12}a^r_{21}$, be the corrected Y-parameters matrix of the FET under reverse cold-FET bias condition. The following expressions represent the parasitic capacitances as a function of the measured Y-parameters:

$$Cpg = Im\{Y^r_{11} + 2Y^r_{12}\}/\omega = \frac{\mathrm{Re}\{Z_C\}}{|Z_C|^2} Im\{Bg\} + \frac{Im\{Z_C\}}{|Z_C|^2} \mathrm{Re}\{Bg\}, \tag{4.126}$$

$$Cpd = Im\{Y^r_{22} + Y^r_{12}\}/\omega = \frac{\mathrm{Re}\{Z_C\}}{|Z_C|^2} Im\{Bd\} + \frac{Im\{Z_C\}}{|Z_C|^2} \mathrm{Re}\{Bd\}, \tag{4.127}$$

with $B_g = (a^r_{22} - 2\Delta a_r)/a^r_{21}$, and $B_d = (a^r_{11} - \Delta a_r)/a^r_{12}$.

Since the calculation of A_g, A_s, A_d , B_g, and B_d terms does not require knowledge of the characteristics of the TRL, TRM, and TRRM structures, the dependence of the parasitic elements extraction on the calibration structures lies entirely on the knowledge of real and imaginary parts of Z_C. However, note that the aforementioned expected sensitivities depend on the real and imaginary parts of the A_g, A_s, A_d , B_g, and B_d terms, which are calculated from measurements of a specific set of calibration structures and the DUT. Thus, the actual impact of $\mathrm{Re}\{Z_C\}$ and $Im\{Z_C\}$ on every FET parasitic element varies in a case-by-case basis.

An example of this sensitivity analysis is presented by using S-parameters measurements of a microwave GaAs FET under both forward and reverse cold-FET bias conditions. The measurements are performed by using a VNA in the frequency range of 0.1–40 GHz, calibrated by using TRM calibration structures. A symmetrical broadband load, of impedance value identical to Z_M, was used as match standard. Thus, for this case, $Z_C = Z_M$. The real and imaginary parts of Z_M are approximated by a resistance in series with an inductor. According to calibration substrate manufacturer, the DC resistance is $R_M = 50\ \Omega$. Meanwhile, its inductance, $L_M = 30$ pH, is determined from its S-parameters corrected using the TRL calibration technique.

The impact of using values of R_M and L_M different from their actual values on the calculation of the FET parasitic elements is showed below. First, every parasitic element was calculated by using values of R_M ranging from

44 to 56 Ω (50 ± 6 Ω) while keeping L_M fixed to 30 pH. Then, the parasitic elements were calculated by using values of L_M ranging from 0 to 60 pH (30 ± 30 pH) while keeping R_M fixed to 50 Ω.

Figure 4.22 shows the calculated parasitic resistances. To better show the impact on parasitic resistances extraction due to the use of R_M and L_M values different from their actual values, data up to 10 GHz are shown. In Figure 4.23, data corresponding to the calculation of the parasitic inductances are shown. No significant frequency-dependence is observed on the data as the value of R_M is varied, compared to the results presented for the parasitic resistances, due to the low impact of the R_M/ω factor at microwave frequencies.

Figure 4.24 shows the calculated parasitic capacitances. It is noted that the extraction of the parasitic capacitances is more sensitive to errors on R_M than to errors on L_M. This is because, in the studied case, the real and imaginary parts of B_g and B_d are of the same order of magnitude and, at microwave frequencies, the ratio R_M/ω is typically several orders of magnitude greater than L_M.

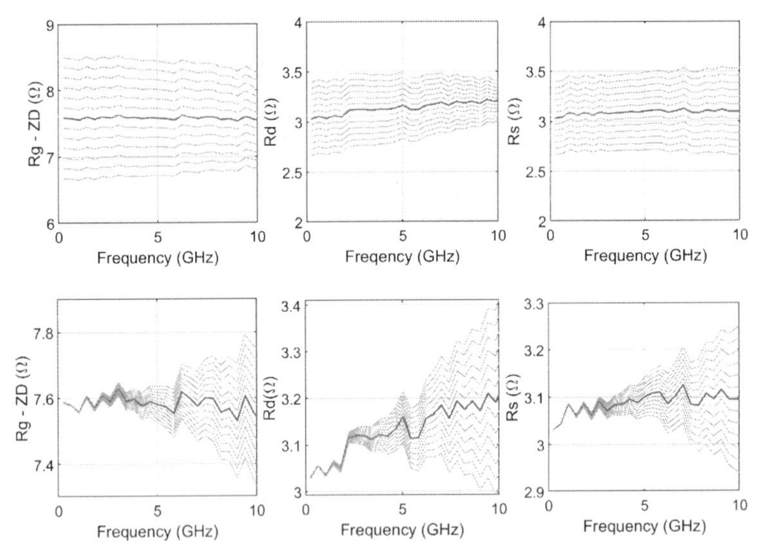

Figure 4.22 (a)R_g, (b) R_d, and (c) R_s calculated by fixing $L_M = 30$ pH and varying R_M from 44 to 56 Ω. (d) R_g, (e) R_d, and (f) R_s calculated by fixing $R_M = 50$ Ω and varying L_M from 0 to 60 pH. Blue traces denote the results obtained by using actual values of $R_M = 50$ Ω and $L_M = 30$ pH. Ref [60]. © 2021 IEEE. Reprinted with permission.

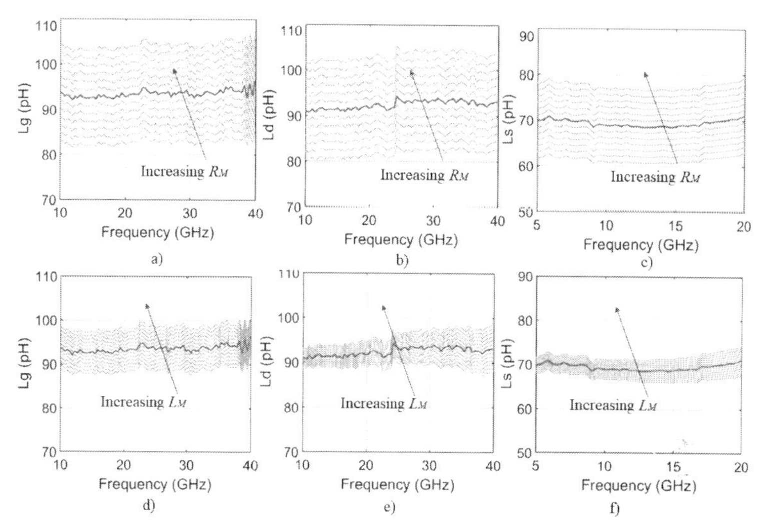

Figure 4.23 (a)L_g, (b) L_d, and (c) L_s calculated by fixing $L_M = 30$ pH and varying R_M from 44 to 56 Ω. (d) L_g, (e) L_d, and (f) L_s calculated by fixing $R_M = 50$ Ω and varying L_M from 0 to 60 pH. Blue traces denote the results obtained by using actual values of $R_M = 50$ Ω and $L_M = 30$ pH. Ref [60]. © 2021 IEEE. Reprinted with permission.

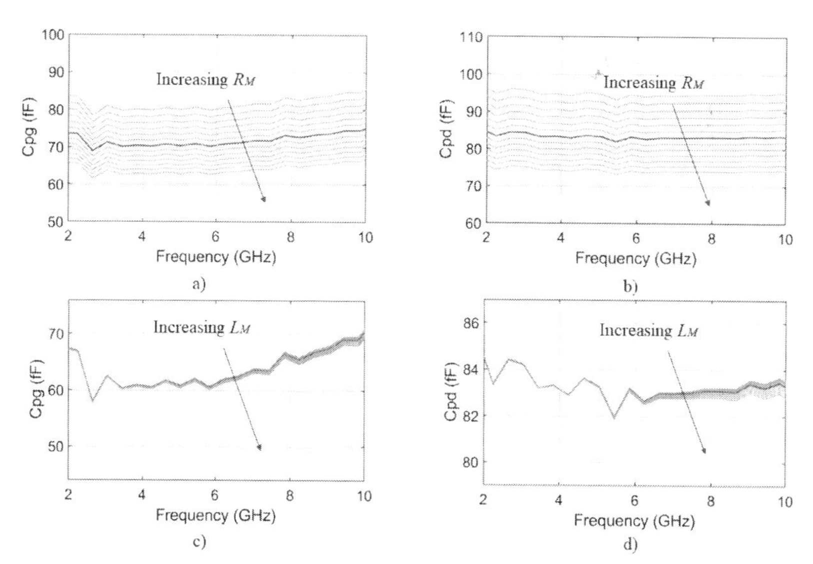

Figure 4.24 (a)C_{pg} and (b) C_{pd} calculated by fixing $L_M = 30$ pH and varying R_M from 44 to 56 Ω. (c) C_{pg} and (d) C_{pd} calculated by fixing $R_M = 50$ Ω and varying L_M from 0 to 60 pH. Blue traces denote the results obtained by using actual values of $R_M = 50$ ω and $L_M = 30$ pH. Ref [60]. © 2021 IEEE. Reprinted with permission.

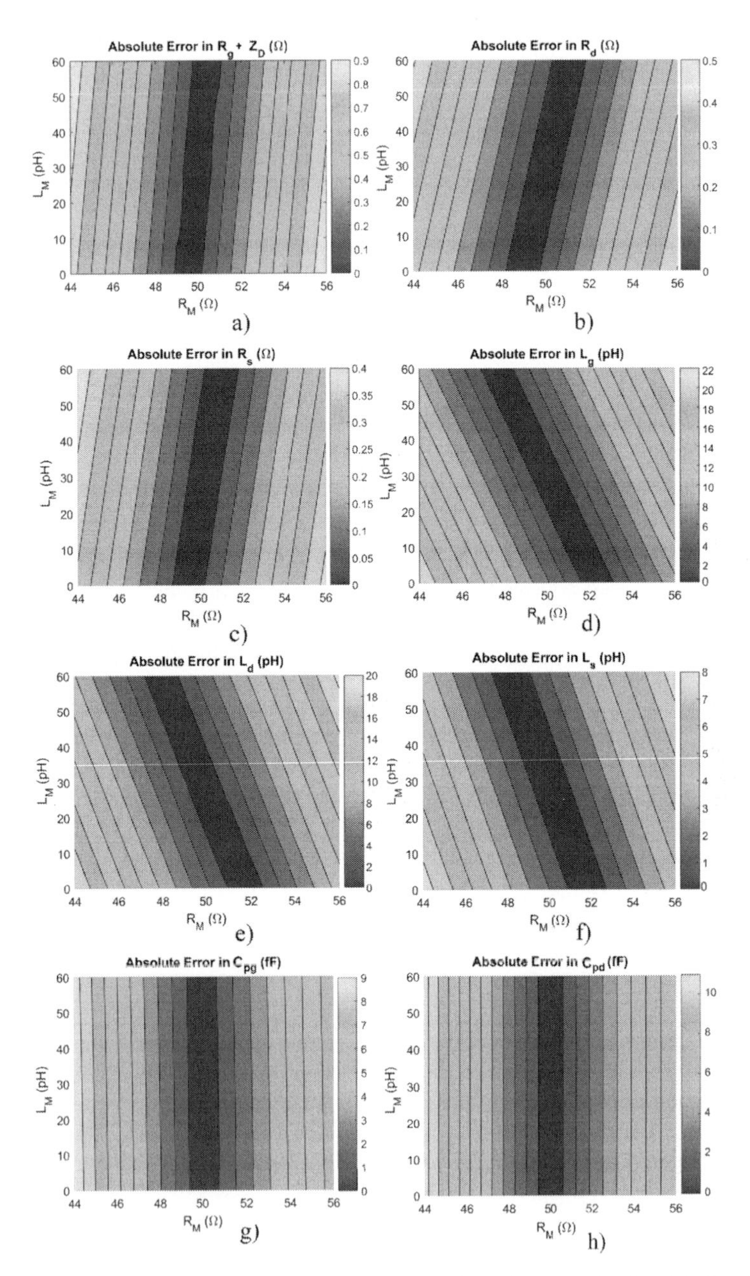

Figure 4.25 Percent error in the extracted FET parasitic elements: (a) R_g, (b) R_d, (c) R_s, (d) L_g, (e) L_d, (f) L_s, (g) C_{pg}, and (h) C_{pd}, caused by varying R_M from 44 to 56 Ω and L_M from 0 to 60 pH. Ref [60]. © 2021 IEEE. Reprinted with permission.

The error in the extracted value of every parasitic element caused by using values of R_M and L_M different from their actual values is shown in Figure 4.25. The error (ϵ) in each parasitic element is evaluated as a percent error as $\epsilon = 100|v_c - v_0|/v_0$, where v_c represents the value of a parasitic element extracted by using the values of v_0 ($50 \pm 6\Omega$) and L_M (30 ± 30 pH) in its calculation; meanwhile, v_0 represents its extracted value calculated by using the actual values of R_M and L_M.

5

Transmission Line Impedance Characterization Methods

Calibration techniques for vector network analyzers (VNA), thru/line-reflect-line (TRL/LRL) [21], line-reflect-match (LRM) [22], and line-reflect-reflect-match (LRRM) [16], use as calibration element a transmission line. In order to accurately refer the device under test (DUT) S-parameters to the measuring system impedance and to set the calibration reference plane to the DUT planes, these calibration techniques require the knowledge of the characteristic impedance, mechanical length, and propagation constant of the transmission line. Usually, the propagation constant of a uniform transmission line, embedded within arbitrary connectors, is determined from the scattering parameter measurements performed on two lines (L-L method) [8], [48], [45] having the same characteristic impedance but different lengths. The L-L method for the line propagation constant, γ, determination may be implemented using either the ABCD matrix [48] or wave cascade matrix [45], [8]. With respect to Z_c, this parameter is not easily determined when the transmission line is embedded between arbitrary connectors [20],[82]. Indirect methods dealing with Z_c determination, based on calibration-independent methods, have been reported in [52] and [93]. Indirect methods for Z_c determination use measurements of both the line propagation constant and the line capacitance. Other works, focused on characteristic impedance determination, use two-tier calibration methods [58], [49], and measurements of γ. Examples of these methods are the ones proposed in [58] and [49]. In [49], the transition in which the line is embedded is modeled with a shunt admittance and ignores the pad losses and pad phase delay. In [58], an analytical method is proposed to calculate Z_c when the line is embedded in between symmetrical and reciprocal connectors. Unfortunately, in [58], the solutions of the equations, which provide Z_c and the elements of the transition matrix E_L used to model the connectors between which the line is embedded, are not sufficiently justified. Many methods have been published for determining the line characteristic impedance of a uniform transmission line [98],[90],[99],[67],[75]. This chapter deals with line characteristics impedance methods.

5.1 The L-L Method to Calculate the Line's Characteristic Impedance of a Uniform Transmission Line [67]

This method assumes that VNA is calibrated at the connector plane. The L-L method uses two uniform transmission lines of different length, but identical impedance and propagation constant embedded in between connectors as depicted in Figure 5.1. The matrix representation in ABCD-parameters of a uniform transmission line of length L_i, characteristic impedance Z_L, and propagation constant γ may be expressed as

$$\mathbf{T_L} = \mathbf{T_Z T_{\lambda_i} T_Z}^{-1}, \tag{5.1}$$

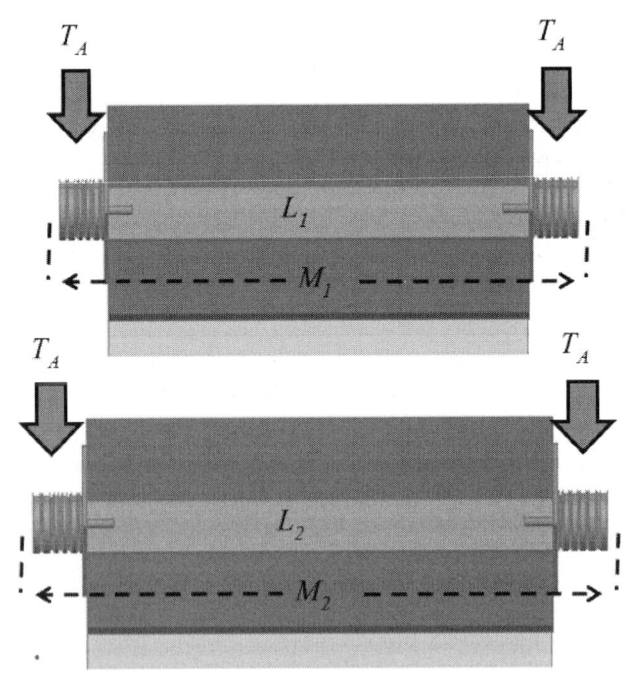

Figure 5.1 Structures utilized for the implementation of L-L method.

where $\mathbf{T_Z}$ and $\mathbf{T_{\lambda_i}}$ are given by

$$\mathbf{T_Z} = \begin{bmatrix} -Z_L & Z_L \\ 1 & 1 \end{bmatrix}, \tag{5.2}$$

$$\mathbf{T_{\lambda_i}} = \begin{bmatrix} \lambda_{Li} & 0 \\ 1 & \lambda_{Li}^{-1} \end{bmatrix}, \tag{5.3}$$

with $\lambda_{Li} = e^{-\gamma l_{Li}}$; $i = 1, 2$. Note that the definition of the matrix $\mathbf{T_Z}$ given in Figure 5.2 differs from that given in [62–64]. The ABCD matrix of the line structures shown in Figure 5.1 may be expressed as

$$\mathbf{M_{Li}} = \mathbf{T_X T_{\lambda_i} T_Y} = \begin{bmatrix} p_{11}^{Li} & P_{12}^{Li} \\ p_{21}^{Li} & p_{22}^{Li} \end{bmatrix}; i = 1, 2, \tag{5.4}$$

where $\mathbf{T_X}$ and $\mathbf{T_Y}$ are defined as follows:

$$\mathbf{T_X} = \mathbf{T_A}\ \mathbf{T_Z} = D_X \begin{pmatrix} \widetilde{A}_X & \widetilde{B}_X \\ \widetilde{C}_X & 1 \end{pmatrix} = \begin{pmatrix} \frac{-Z_L+\widetilde{a}_{12}}{Z_L\widetilde{a}_{21}+1} & \frac{Z_L+\widetilde{a}_{12}}{Z_L\widetilde{a}_{21}+1} \\ \frac{-Z_L\widetilde{a}_{21}+1}{Z_L\widetilde{a}_{21}+1} & 1 \end{pmatrix}, \tag{5.5}$$

$$\mathbf{T_Y} = \mathbf{T_Z^{-1}}\ \mathbf{T_B} = D_Y \begin{pmatrix} \widetilde{A}_Y & \widetilde{B}_Y \\ \widetilde{C}_Y & 1 \end{pmatrix} = \frac{Z_L + \widetilde{a}_{12}}{2Z_L det(T_A)} \begin{pmatrix} \frac{Z_L\widetilde{a}_{21}-1}{Z_L+\widetilde{a}_{12}} & \frac{Z_L-\widetilde{a}_{12}}{Z_L+\widetilde{a}_{12}} \\ \frac{Z_L\widetilde{a}_{21}+1}{Z_L+\widetilde{a}_{12}} & 1 \end{pmatrix}, \tag{5.6}$$

In eqn (5.5) and (5.6), matrices $\mathbf{T_A}$ and $\mathbf{T_B}$ represent the ABCD-parameters of the connectors or transitions from coplanar to microstrip at ports 1 and 2 of the VNA, in which the transmission lines are embedded. The connectors, which are assumed to be identical and symmetrical ($a_{11} = a_{22}$) two-port networks, are defined as

$$\mathbf{T_A} = \begin{bmatrix} a_{11} & a_{12} \\ a_{21} & a_{11} \end{bmatrix}. \tag{5.7}$$

$$\mathbf{T_B} = \frac{1}{\det(\mathbf{T_A})} \begin{bmatrix} a_{11} & a_{12} \\ a_{21} & a_{11} \end{bmatrix}. \tag{5.8}$$

In eqn (5.5) and (5.6), terms denoted as $\widetilde{a}_{ij}$; $i, j = 1, 2$ represent the elements of the matrices $\mathbf{T_A}$ and $\mathbf{T_B}$ normalized to a_{11}. Since in the proposed procedure, the connectors or transitions (coplanar to microstrip) do not necessarily have to be reciprocal, the condition $\det(\mathbf{T_A}) = 1$ is not required. By using the L-L method, two different procedures for determining the line characteristic impedance Z_L are presented next.

5.2 Procedure I: Calculation of Characteristic Impedance Using Both the Partial Knowledge of T_X Matrix and Line Propagation Constant γ

From the definition of the elements of $\mathbf{T_X}$ given in eqn (5.5), the following expression for determining the line characteristic impedance may be derived:

$$Z_L = \frac{\widetilde{B_X} - \widetilde{A_X}}{1 + \widetilde{C_X}}. \tag{5.9}$$

For determining the value of Z_L, it is necessary to know the value of the terms $\widetilde{A_X}$,$\widetilde{B_X}$, and $\widetilde{C_X}$. In the classical TRL calibration technique, $\widetilde{C_X}$ is calculated using the measurement of a symmetrical reflecting load. Meanwhile, in the proposed method, this term is determined from the reciprocity of the transitions. Solving eqn (5.4) for $\mathbf{T_Y}$ and developing the resulting expression, the following definitions for $\widetilde{A_Y}$, $\widetilde{B_Y}$, $\widetilde{C_Y}$, and $\widetilde{D_X D_Y}$ are obtained

$$\mathbf{D_X} D_Y \begin{pmatrix} \widetilde{A}_Y & \widetilde{B}_Y \\ \widetilde{C}_Y & 1 \end{pmatrix}$$

$$= \frac{-p_{12}^{Li} + \frac{\widetilde{A_X}}{\widetilde{C_X}} p_{22}^{Li}}{\frac{\widetilde{A_X}}{\widetilde{C_X}} - \widetilde{B_X}} \lambda_L i \begin{pmatrix} \frac{\frac{1}{\widetilde{C_X}} p_{11}^{Li} - \widetilde{B_X} p_{21}^{Li}}{-p_{12}^{Li} + \frac{\widetilde{A_X}}{\widetilde{C_X}} p_{21}^{Li}} \frac{1}{\lambda_{Li}^2} & \frac{\frac{1}{\widetilde{C_X}} p_{12}^{Li} - \widetilde{B_X} p_{22}^{Li}}{-p_{12}^{Li} + \frac{\widetilde{A_X}}{\widetilde{C_X}} p_{21}^{Li}} \frac{1}{\lambda_{Li}^2} \\ \frac{-p_{11}^{Li} - \frac{\widetilde{A_X}}{\widetilde{C_X}} p_{21}^{Li}}{-p_{12}^{Li} + \frac{\widetilde{A_X}}{\widetilde{C_X}} p_{21}^{Li}} & 1 \end{pmatrix}. \tag{5.10}$$

Note from eqn (5.5) and (5.6) that $\widetilde{A_Y}$ may also be expressed in terms of $\widetilde{B_X}$ and $\widetilde{C_X}$ as

$$\widetilde{A_Y} = \frac{\widetilde{C_X}}{\widetilde{B_X}}. \tag{5.11}$$

Then, equating the expression for $\widetilde{A_Y}$ given in eqn (5.10) and (5.11), and using the measurement of one of the two transmission lines, the value of $\widetilde{C_X}$

may be calculated

$$\widetilde{C_X} = \frac{1}{\lambda_{Li}} \sqrt{\widetilde{B_X} \frac{\widetilde{B_X} p_{21}^{Li} - p_{11}^{Li}}{\frac{\widetilde{A_X}}{\widetilde{C_X}} p_{21}^{Li} - p_{12}^{Li}}}. \tag{5.12}$$

Once $\widetilde{A_X}$, $\widetilde{B_X}$, and $\widetilde{C_X}$ are known, the characteristic impedance of the transmission lines is determined from eqn (5.9). It can be concluded from eqn (5.9) and (5.12) that the calculation of the line characteristic impedance depends on the value of the propagation constant and the length of only one transmission line. Moreover, as in the TRL [72],[71],[21] the solution for the terms $\widetilde{A_X}$, $\widetilde{B_X}$, and $\widetilde{C_X}$ (in the proposed method, necessary for determining Z_L) becomes ill-conditioned when the phase shift of the pair of lines (angle of λ_Δ) is a multiple of $180°$. Thus, the lengths of the transmission lines have to be correctly chosen according to [74].

5.2.1 Calculation of $\frac{A_X}{C_X}$, $\widetilde{B_X}$ and the propagation constant γ [74]

The procedure for determining the propagation constant is based on the theory of the TRL method [72], [71], [21]. Calculating $\mathbf{M_{21}} = \mathbf{M_2 M_1^{-1}}$ one has

$$\mathbf{M_{21}} = \mathbf{T_X T_{\lambda_\Delta} T_X^{-1}} = \begin{bmatrix} m_{11} & m_{12} \\ m_{21} & m_{22} \end{bmatrix}, \tag{5.13}$$

$$\mathbf{T}_{\lambda_\Delta} = \begin{bmatrix} \lambda_\Delta & 0 \\ 1 & \lambda_\Delta^{-1} \end{bmatrix}, \tag{5.14}$$

and $\Delta l = l_{L_2} - l_{L_1}$. Then solving eqn (5.13) for $\mathbf{T}_{\lambda_\Delta}$ and developing the resulting expression, the following equations are obtained:

$$\overline{\frac{A_X}{C_X}}, \overline{B_X} = \frac{(m_{11} - m_{22}) \pm \sqrt{(m_{11} - m_{22})^2 - 4(m_{12} m_{21})}}{2m_{21}}, \tag{5.15}$$

$$\lambda_L = \frac{\overline{\frac{A_X}{C_X}} m_{11} + m_{21} - \overline{\frac{A_X}{C_X}}\, \overline{B_X} m_{21} - \overline{B_X} m_{22}}{\overline{\frac{A_X}{C_X}} - \overline{B_X}}. \tag{5.16}$$

The propagation constant is then calculated as

$$\gamma = \frac{\ln \lambda_\Delta}{\Delta l}. \tag{5.17}$$

By choosing the appropriate root in eqn (5.15), the values of $\widetilde{A_X}/\widetilde{C_X}$ and $\widetilde{B_X}$ are determined and utilized to determine the line characteristic impedance from eqn (5.9).

5.3 Procedure II: Method to Calculate the Line's Characteristic Impedance [98], [99]

When a uniform transmission line of arbitrary length and arbitrary characteristic impedance is embedded in between connectors or coplanar to waveguide transitions, as shown in Figure 5.2, it is difficult- if not impossible- [20], [82] to determine its characteristic impedance directly from a single set of S-parameter measurements. According to [20], the structure shown in Figure 5.2 consists of a uniform transmission line and two connectors referred to as A and B, represented using the formalism of the ABCD transmission matrix as $\mathbf{T_{Li}}$, $\mathbf{T_A}$, and $\mathbf{T_B}$ matrices, respectively. Using the ABCD matrix formalism, the equivalent matrix $\mathbf{M}$ of the structure shown in Figure 5.2 is equal to the product of the three individual matrices as follows:

$$\mathbf{M} = \mathbf{T_A}\mathbf{T_{Li}}\mathbf{T_B}, \tag{5.18}$$

with

$$\mathbf{T_A} = \begin{bmatrix} a_{11} & a_{12} \\ a_{21} & a_{22} \end{bmatrix}, \tag{5.19}$$

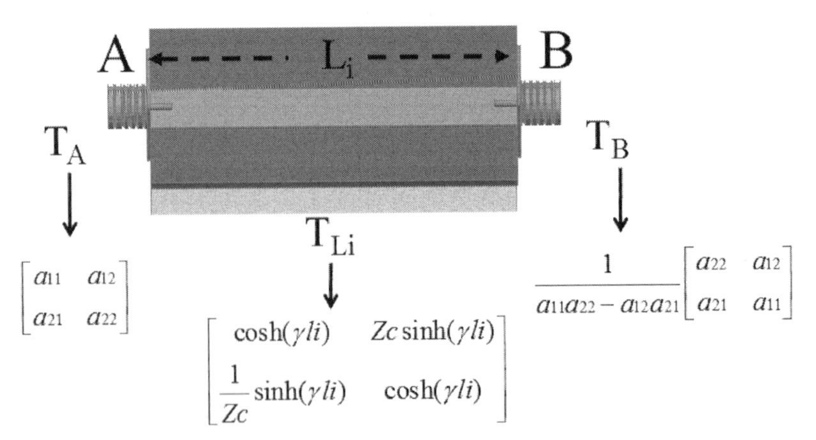

Figure 5.2 Uniform transmission line of arbitrary length and arbitrary characteristic impedance embedded in coaxial connectors.

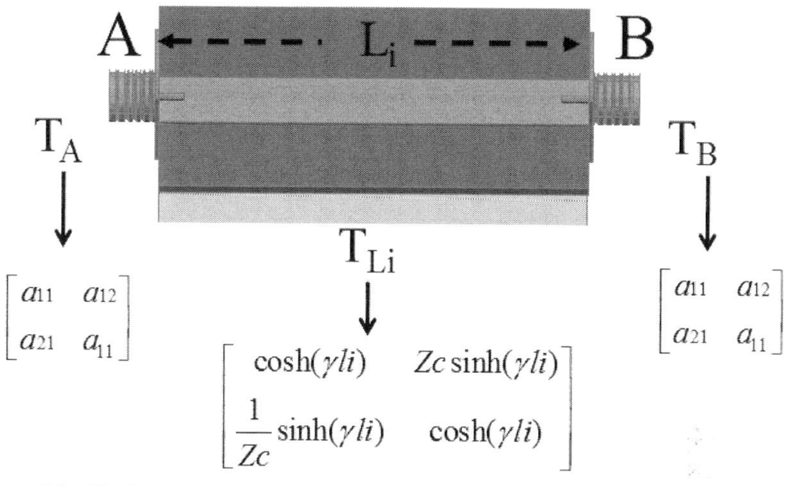

Figure 5.3 Uniform transmission line of arbitrary length and arbitrary characteristic impedance embedded in symmetrical and reciprocal coaxial connectors.

$$\mathbf{T_B} = \frac{1}{a_{11}a_{22} - a_{12}a_{21}} \begin{bmatrix} a_{11} & a_{12} \\ a_{21} & a_{22} \end{bmatrix}; \tag{5.20}$$

and

$$\mathbf{T_{Li}} = \begin{bmatrix} \cosh(\gamma Li) & Z_c\sinh(\gamma Li) \\ \frac{1}{Z_c}\sinh(\gamma)Li & \cosh(\gamma)Li \end{bmatrix}. \tag{5.21}$$

where $\mathbf{T_{Li}}$ is the ABCD matrix for the uniform transmission line, γ is the line propagation constant, Z_c is the line characteristic impedance, and l_i is the length. Under the assumption that the connectors A and B are identical ($B = A$), symmetrical $a_{22} = a_{11}$, and reciprocal $(a_{11}a_{22} - a_{21}a_{12}) = 1$ as shown in Figure 5.3, the equivalent matrix $\mathbf{M}$ of the structure can be expressed as follows:

$$\begin{aligned} \mathbf{M} &= \mathbf{T_A T_{Li} T_B} \\ &= \begin{bmatrix} a_{11} & a_{12} \\ a_{21} & a_{22} \end{bmatrix} \cdot \begin{bmatrix} \cosh(\gamma Li) & Z_c\sinh(\gamma Li) \\ \frac{1}{Z_c}\sinh(\gamma)Li & \cosh(\gamma)Li \end{bmatrix} \cdot \begin{bmatrix} a_{11} & a_{12} \\ a_{21} & a_{22} \end{bmatrix}. \end{aligned} \tag{5.22}$$

The method proposed to calculate Z_c requires S-parameter measurements performed on two uniform transmission lines having the same characteristic impedance and propagation constant but different lengths as shown in

Figure 5.1. The ABCD matrices $\mathbf{M_1}$ and $\mathbf{M_2}$ resulting from the measurements of lines l_1 and l_2 are expressed as

$$\mathbf{M_1} = \mathbf{T_A T_{L1} T_A}, \tag{5.23}$$

$$\mathbf{M_2} = \mathbf{T_A T_{L2} T_A}, \tag{5.24}$$

where

$$\mathbf{M_1} = \begin{bmatrix} m_{11} & m_{12} \\ m_{21} & m_{22} \end{bmatrix}, \tag{5.25}$$

and

$$\mathbf{M_2} = \begin{bmatrix} p_{11} & p_{12} \\ p_{21} & p_{22} \end{bmatrix}. \tag{5.26}$$

In order to find an analytical expression to determine the line's characteristic impedance, eqn (5.23) and (5.24) are written as

$$\mathbf{M_1 T_A}^{-1} = \mathbf{T_A T_{L1}}, \tag{5.27}$$

$$\mathbf{M_2 T_A}^{-1} = \mathbf{T_A T_{L2}}. \tag{5.28}$$

Then, taking advantage of the symmetry property of the structure shown in Figure (5.3), $\mathbf{T_B} = \mathbf{T_A}$, it is easy to conclude from eqn (5.23) and (5.24) that $m_{11} = m_{22}$ and $p_{11} = p_{22}$. Using these results and writing out eqn (5.27), it follows that:

$$m_{11} - \cosh(\gamma l_1) = m_{12}\frac{a_{21}}{a_{11}} + \frac{1}{Z_c}\sinh(\gamma l_1)\frac{a_{12}}{a_{11}}, \tag{5.29}$$

$$m_{1_1} - \cosh(\gamma l_1) = m_{21}\frac{a_{12}}{a_{11}} + Z_c\sinh(\gamma l_1)\frac{a_{21}}{a_{11}}, \tag{5.30}$$

$$m_{12} = (m_{11} + \cosh(\gamma l_1))\frac{a_{12}}{a_{11}} + Z_c\sinh(\gamma l_1), \tag{5.31}$$

$$m_{21} = (m_{11} + \cosh(\gamma l_1))\frac{a_{21}}{a_{11}} + \frac{1}{Z_c}\sinh(\gamma l_1). \tag{5.32}$$

In the same way, but now writing out eqn (5.28), one has

$$p_{11} - \cosh(\gamma l_2) = p_{12}\frac{a_{21}}{a_{11}} + \frac{1}{Z_c}\sinh(\gamma l_2)\frac{a_{12}}{a_{11}}, \tag{5.33}$$

$$p_{11} - \cosh(\gamma l_2) = p_{21}\frac{a_{12}}{a_{11}} + Z_c\sinh(\gamma l_2)\frac{a_{21}}{a_{11}}, \tag{5.34}$$

$$p_{12} = (p_{11} + \cosh(\gamma l_2))\frac{a_{12}}{a_{11}} + Z_c \sinh(\gamma l_2), \tag{5.35}$$

$$p_{21} = (p_{11} + \cosh(\gamma l_2))\frac{a_{21}}{a_{11}} + \frac{1}{Z_c}\sinh(\gamma l_2). \tag{5.36}$$

Then, a set of two simultaneous equations is obtained from eqn (5.23) and (5.24). Moreover, the first set of simultaneous equations, eqn (5.31) and (5.35), can be expressed in matrix form as

$$\begin{bmatrix} m_{12} \\ p_{12} \end{bmatrix} = K \begin{bmatrix} \frac{a_{12}}{a_{11}} \\ Z_c \end{bmatrix}. \tag{5.37}$$

And the second set of simultaneous equations, eqn (5.32) and (5.36) can be expressed in matrix form as

$$\begin{bmatrix} m_{21} \\ p_{21} \end{bmatrix} = K \begin{bmatrix} \frac{a_{21}}{a_{11}} \\ \frac{1}{Z_c} \end{bmatrix}, \tag{5.38}$$

where

$$\mathbf{K} = \begin{bmatrix} m_{11} + \cosh(\gamma l_1) & \sinh(\gamma l_1) \\ p_{21} + \cosh(\gamma l_2) & \sinh(\gamma l_1) \end{bmatrix}. \tag{5.39}$$

Since eqn (5.37) and (5.38) have the same **K** matrix, they can be grouped as

$$\mathbf{M_x} = \mathbf{KB} \tag{5.40}$$

with

$$\mathbf{M_x} = \begin{bmatrix} m_{11} & m_{12} \\ p_{21} & p_{12} \end{bmatrix} \tag{5.41}$$

and

$$\mathbf{B} = \begin{bmatrix} \frac{a_{11}}{a_{11}} & \frac{a_{12}}{a_{11}} \\ \frac{1}{Z_c} & Z_C \end{bmatrix}. \tag{5.42}$$

Solving eqn (5.40) for B, one has:

$$\mathbf{B} = \mathbf{K}^{-1}\mathbf{M_x}. \tag{5.43}$$

Finally, an analytical expression to calculate Z_c is obtained from eqn (5.43) as:

$$Z_C = \frac{p_{12}(\cosh(\gamma l_1)) + m_{12}(\cosh(\gamma l_2)) + p_{11}}{\det(K)}, \tag{5.44}$$

where $\det(K)$ is the value of the determinant of K matrix, expressed as

$$\det(\mathbf{K}) = (m_{11} + \cosh(\gamma l_1))\sinh(\gamma l_2) - (p_{11} + \cosh(\gamma l_2))\sinh(\gamma l_1). \tag{5.45}$$

It is important to comment that the analytical expression to calculate Z_c depends only on the length of lines l_1, l_2 and the wave propagation constant γ. This expression is different from the one reported in [90] and [58]. The proposed method to calculate the line's characteristic impedance requires the knowledge of the propagation constant, which can be determined using the method proposed in [74]. In order to show the usefulness of the L-L method for line characteristic impedance determination, two examples are given. Both examples use a VNA calibrated in the frequency range of 0.01–20 GHz at the end of the probe tips (coplanar devices) or at the end of the coaxial connectors (coaxial connectorized lines).

The first example uses two transmission lines embedded in pads, fabricated in a silicon substrate. The lengths of these two transmission lines are 540 and 600 μm. The phase shift of this pair of transmission lines was calculated from eqn (5.16) as $15°$ at the highest frequency.

The real and imaginary parts of the characteristic impedance of these transmission lines, calculated using the proposed method and the L-L method (procedure I and procedure II), are shown in Figure 5.4 and 5.5, respectively. A high correlation between the real and imaginary parts of ZL calculated using both procedures is achieved. This suggests that the used values of the mechanical lengths of both used transmission lines are nearly approximated to their actual values. Moreover, since the phase shift of the pair of lines is less than 180° over the whole frequency range of interest, Z_L is calculated in a continuous manner.

In the second example, microstrip transmission lines fabricated in a Rogers RT/Duroid substrate ϵ_r=2.33, and embedded in the coplanar waveguide to microstrip transitions, were used for implementing the proposed method. The lengths of these transmission lines are 1.27 and 2.54 cm.

Procedure I and procedure II were implemented and Z_L was determined. Figures 5.6 and 5.7 show the calculated real and imaginary parts of Z_L, respectively. It can be noted that in both procedures, Z_L is not calculated in a continuous manner over the whole frequency range of interest; in both real and imaginary parts, spikes are observed. Figure 5.8, shows the phase shift of the used pair of microstrip transmission lines. It can be noted that the spikes in Z_L occur at frequencies where the angle of equals $180°$ and $360°$.

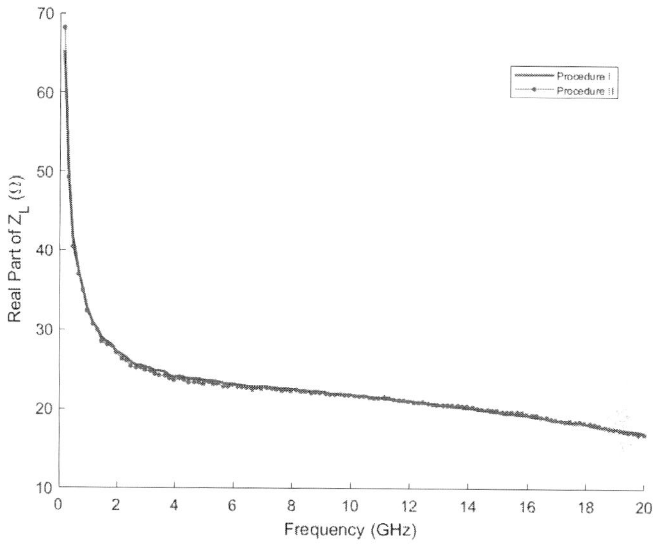

Figure 5.4 The real part of Z_L of a transmission line calculated using two lines with phase shift less than 180°.

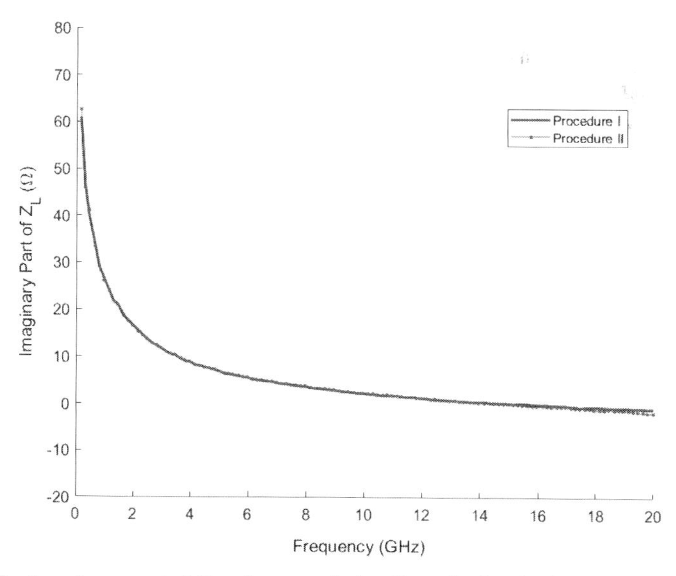

Figure 5.5 Imaginary part of Z_L of a transmission line calculated using two lines with phase shift less than 180°.

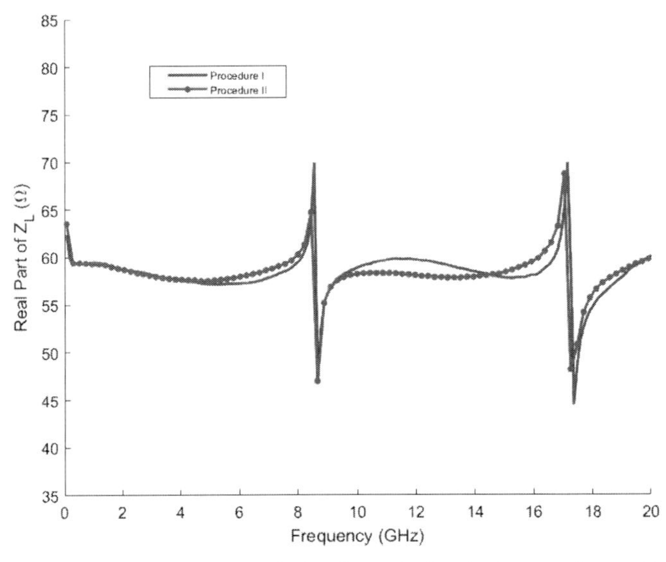

Figure 5.6 Real part of Z_L of a transmission line calculated using two lines with phase shift greater than 180° using two lines presenting phase shift of 180° and 360° at 8.8 and 17.6 GHz, respectively.

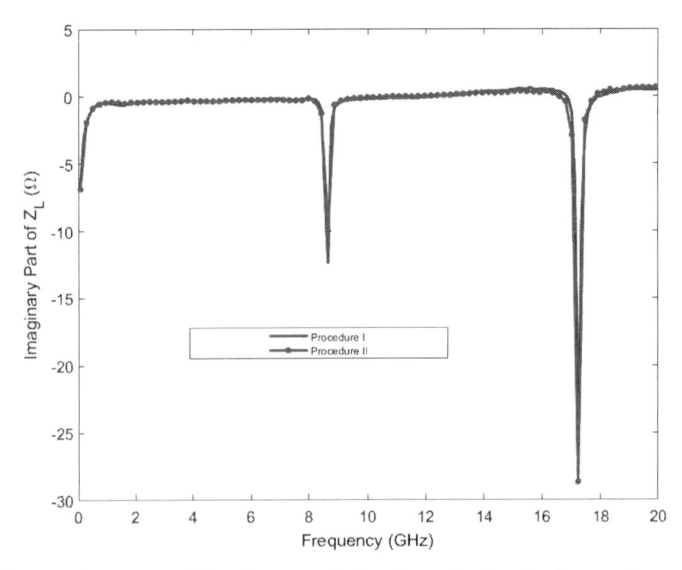

Figure 5.7 Imaginary part of ZL of a transmission line calculated using two lines with phase shift greater than 180°.

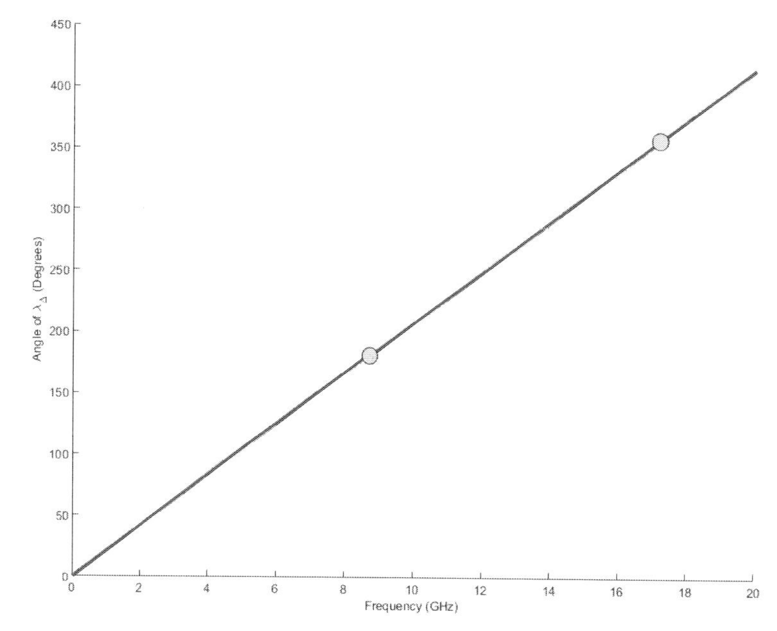

Figure 5.8 Imaginary part of ZL of a transmission line calculated using two lines with phase shift greater than 180°.

5.4 Method to Calculate the Line's Characteristic Impedance Using Calibration Comparison [67]

The calibration comparison methods, which compare two calibration techniques, are implemented using the eight-error term model to represent an uncalibrated VNA. An uncalibrated two-port VNA (Figure 5.9), in which the switching and isolation errors are accounted for, measures the product of three

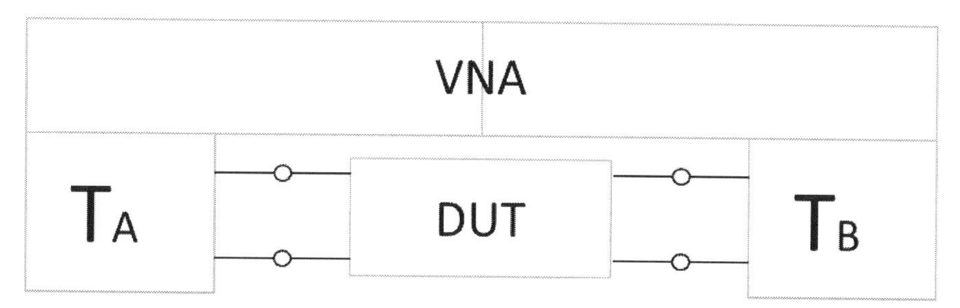

Figure 5.9 Uncalibrated measurement using a two-port VNA (switching terms and isolation errors are previously corrected). Ref [67]. © 2018 IEEE. Reprinted with permission.

matrices

$$\mathbf{M_D} = \mathbf{T_A T_D T_B}. \tag{5.46}$$

$\mathbf{M_D}$ and $\mathbf{T_D}$ are matrices describing, respectively, the measured and the actual behavior of a DUT. $\mathbf{T_A}$ and $\mathbf{T_B}$ are matrices that model the error boxes in ports 1 and 2 of the VNA. The matrix $\mathbf{T_D}$ may be expressed as a function of the matrix $\mathbf{M_D}$ as follows:

$$\mathbf{T_D} = \mathbf{T_A^{-1} M_D T_B^{-1}} = \frac{1}{D_X D_Y} \mathbf{T_Q} \widetilde{\mathbf{T_X^{-1}}} \mathbf{M_D} \widetilde{\mathbf{T_Y^{-1}}} \mathbf{T_Q^{-1}}, \tag{5.47}$$

where

$$D_X \widetilde{\mathbf{T_X}} = \mathbf{T_A T_Q}, \tag{5.48}$$

$$D_X \widetilde{\mathbf{T_Y}} = \mathbf{T_Q^{-1} T_B}. \tag{5.49}$$

The matrices $D_X \widetilde{\mathbf{T_X}}$ and $D_Y \widetilde{\mathbf{T_Y}}$ are modified versions of $\mathbf{T_A}$ and $\mathbf{T_B}$, respectively, and $\mathbf{T_Q}$ is a matrix dependent on the impedance of the one calibration standards (the line in a TRL calibration or the load in a TRM and TRRM calibration). The elements of matrices $\widetilde{\mathbf{T_X}}$ and $\widetilde{\mathbf{T_Y}}$ along with the product $D_x D_Y$ are determined by using some calibration technique, e.g., TRL, TRM, and TRRM [67]. Moreover, in order to correctly calculate $\mathbf{T_D}$, the elements of $\mathbf{T_Q}$ have to be known prior to the calibration.

Thus, the value of the elements of matrices $\widetilde{\mathbf{T_X}}$, $\widetilde{\mathbf{T_Y}}$, and $\mathbf{T_Q}$ along with the term $D_X D_Y$ depends on the calibration technique and also on the parameters (T or ABCD) used to implement it. The matrices $\mathbf{T_A}$ and $\mathbf{T_B}$ depend on the measurement setup and on the parameters used to implement it; they do not depend on the calibration technique.

Therefore, to be consistent, the matrix $\mathbf{T_A} = D_X \widetilde{\mathbf{T_X}} \mathbf{T_Q^{-1}}$ corresponding to two different calibration techniques has to be identical. The same holds for matrix $\mathbf{T_B} = D_Y \widetilde{\mathbf{T_Y}} \mathbf{T_Q}$. The following expressions are derived by combining the calibration terms obtained from the implementation of two calibration techniques "i" and "j".

$$\mathbf{T_{Qj}^{-1} T_{Qi}} = \frac{D_{Xi}}{D_{Xj}} X = \frac{D_{Xi}}{D_{Xj}} \begin{bmatrix} X_{11} & X_{12} \\ X_{21} & X_{22} \end{bmatrix}, \tag{5.50}$$

$$\mathbf{T_{Qj}^{-1} T_{Qi}} = \frac{D_{Yi}}{D_{Yj}} Y = \frac{D_{Yj}}{D_{Yi}} \begin{bmatrix} Y_{11} & Y_{12} \\ Y_{21} & Y_{22} \end{bmatrix}, \tag{5.51}$$

where $\mathbf{X} = \widetilde{\mathbf{T_{X_j}^{-1}}} \widetilde{\mathbf{T_{X_i}}}$ and $\mathbf{Y} = \widetilde{\mathbf{T_{Y_j}^{-1}}} \widetilde{\mathbf{T_{Y_i}}}$. The proposed method uses calibration terms obtained from the use of calibration techniques implemented using either T- or ABCD-parameters. When the calibration techniques are implemented using T-parameters, the matrix $\mathbf{T_{Q_k}}, k = i, j$, is defined as [74]

$$\mathbf{T_{Q_k}} = \begin{bmatrix} 1 & \Gamma_k \\ \Gamma_k & 1 \end{bmatrix}. \tag{5.52}$$

Here, $\Gamma_k = \frac{Z_k - Z_0}{Z_k + Z_0}$. Meanwhile, when the calibration techniques are implemented using ABCD-parameters, $\mathbf{T_{Q_k}}, k = i, j$, is defined as [65]

$$\mathbf{T_{Qk}} = \begin{bmatrix} -Z_k & Z_k \\ 1 & 1 \end{bmatrix}. \tag{5.53}$$

In eqn (5.52) and (5.53) Z_k represents the line impedance in the TRL calibration and the load impedance in the TRM or TRRM calibrations. The ratio $\frac{Z_i}{Z_j}$ given by eqn (5.54) may be calculated from either eqn (5.50) or eqn (5.51). Then developing $\mathbf{T_{Qj}^{-1}T_{Qi}}$ and equating term-by-term the elements of the resulting matrix with the elements of the matrix at the right side of either eqn (5.50) or eqn (5.51), the following expression may be determined:

$$\frac{Z_j}{Z_i} = \frac{1+h}{1-h}, \tag{5.54}$$

where

$$h = s\frac{X_{21}}{X_{11}} = s\frac{X_{12}}{X_{12}} = s\frac{Y_{21}}{Y_{11}} = s\frac{Y_{12}}{Y_{22}} \tag{5.55}$$

$$s = \pm 1. \tag{5.56}$$

The "+" and "-" signs in eqn (5.56) correspond, respectively, to the use of calibration techniques implemented using T- or ABCD-parameters in the calculation of h. As shown in eqn (5.54) to determine the impedance Z_i, it is mandatory to know the impedance Z_j and vice versa. Therefore, to determine the impedance of a line used in the TRL calibration, it is necessary to know the load impedance used in a TRM or TRRM calibration. Moreover, when the same calibration technique is implemented using two different sets of calibration standards (different substrates), it is mandatory to know the line impedance (for TRL) or the load impedance (for TRM or TRRM) corresponding to one of the two sets of calibration standards.

The proposed method was used to determine the characteristic impedance of a transmission line by comparing the TRL calibration with the TRM and TRRM calibrations. TRL, TRM, and TRRM calibration techniques were performed using calibration standards included in the CS5 calibration substrate from GGB Industries. All of the implemented calibration techniques were performed using ABCD-parameters [63]. The reference plane of all of the calibration techniques was set at the center of a 100-μm transmission line, which was used as a thru standard. Offset reflecting loads (open circuits and short circuits) and offset matched loads were used, respectively, as reflect and match standards.

In order to demonstrate the usefulness of the proposed method, the characteristic impedance (Z_L) of a 500-μm transmission line, used as standard in the TRL calibration, was determined from the impedance of the load used as match standard in the TRM and TRRM calibrations. The real part of the load was assumed to be 50 Ω over the whole frequency range of interest and its imaginary part was estimated using the procedure reported in [42]. The accuracy in the determination of the line impedance using the proposed method was assessed by comparing the real and imaginary parts of the calculated line impedance with those calculated from the line propagation constant and capacitance [93] (NIST Method). Figures 5.10 and 5.11 show the real and imaginary parts of the line impedance calculated from

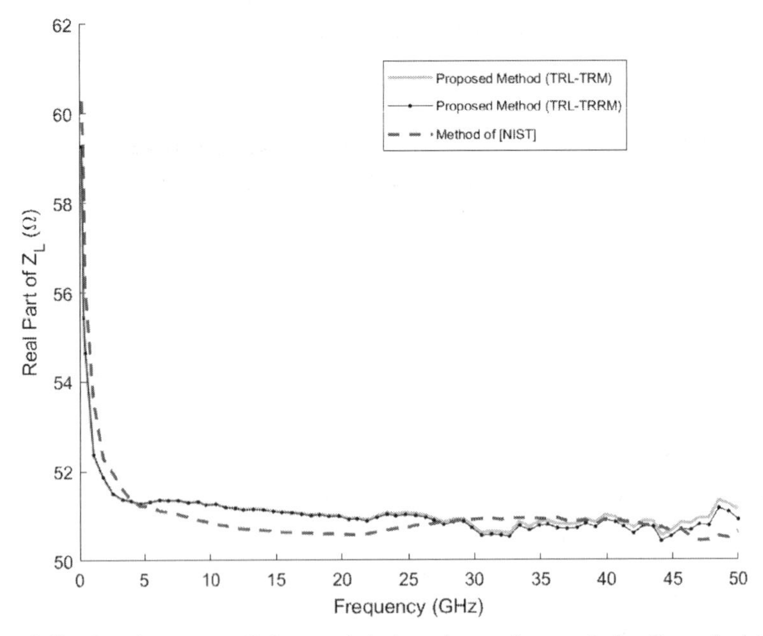

Figure 5.10 Imaginary part of characteristic impedance of transmission line calculated by comparing the TRL calibration with TRM and TRRRM calibrations. Ref [67]. © 2018 IEEE. Reprinted with permission.

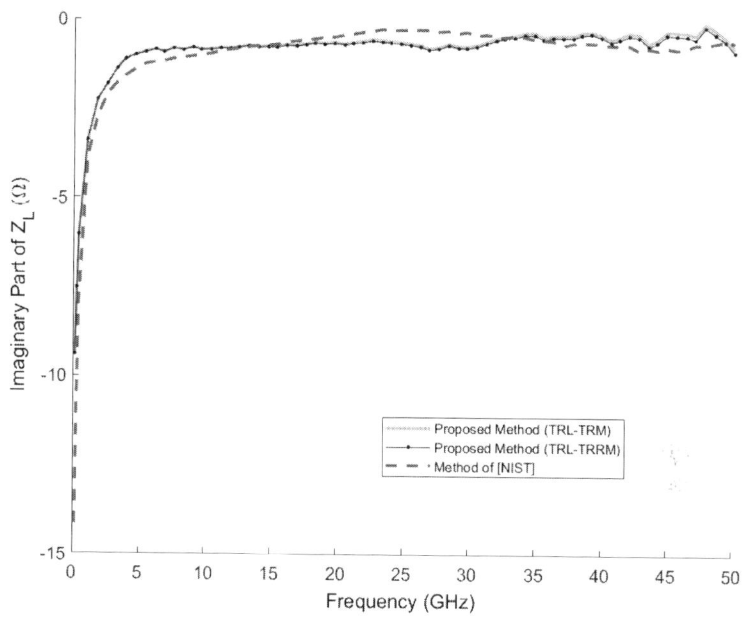

Figure 5.11 Imaginary part of characteristic impedance of transmission line calculated by comparing the TRL calibration with TRM and TRRRM calibrations. Ref [67]. © 2018 IEEE. Reprinted with permission.

eqn ((5.54)) and ((5.55)), comparing the TRL calibration with the TRM and TRRM calibrations. High correlation between the results obtained using the proposed method with the NIST method [93] validates the usefulness of the proposed method.

5.5 Method to Calculate the Line's Characteristic Impedance Using an Uncalibrated Vector Network Analyzer [75]

The method for determining the line characteristic impedance uses the eight-error terms model for modeling a VNA and is based on the fact that the values of the error terms are unique and independent of the calibration structures used to determine them. The main feature of this method relies on the use of raw S-parameter measurements of a thru-line and thru-load combinations to determine the line's characteristic impedance instead of the use of a pair of two-port calibration techniques.

The proposed method for calculating the line characteristic impedance uses an uncalibrated VNA, whose errors terms are represented by matrices $\mathbf{T_A}$ and $\mathbf{T_B}$, as reported in Figure 5.12, and their derivation starts with the measurement of a thru. This measurement is done with an uncalibrated VNA and is defined by matrix $\mathbf{M_{Thru}}$ that is computed using the eight-term error model and the ABCD-parameters matrix formalism. Then as in [67], the equivalent matrix $\mathbf{M_{Thru}}$ of the structure shown in Figure 5.12 is given by

$$\mathbf{M_{Thru}} = \mathbf{T_A I T_B} = \mathbf{T_A T_{Zi} T_{Zi}^{-1} T_B} = \mathbf{T_{Xi} T_{Yi}}, \tag{5.57}$$

where

$$\mathbf{I} = \mathbf{T_{Zi} T_{Zi}^{-1}}, \tag{5.58}$$

and $\mathbf{T_{Zi}}$ is given as

$$\mathbf{T_{Qk}} = \begin{bmatrix} -Z_i & Z_i \\ 1 & 1 \end{bmatrix}. \tag{5.59}$$

In eqn (5.59), Z_i is defined as the impedance of the calibration structures used in a given calibration technique, typically a line or a load as reported in Figure 5.12. The matrices $\mathbf{T_{Xi}}$ and $\mathbf{T_{Yi}}$ are nonunique since they depend on

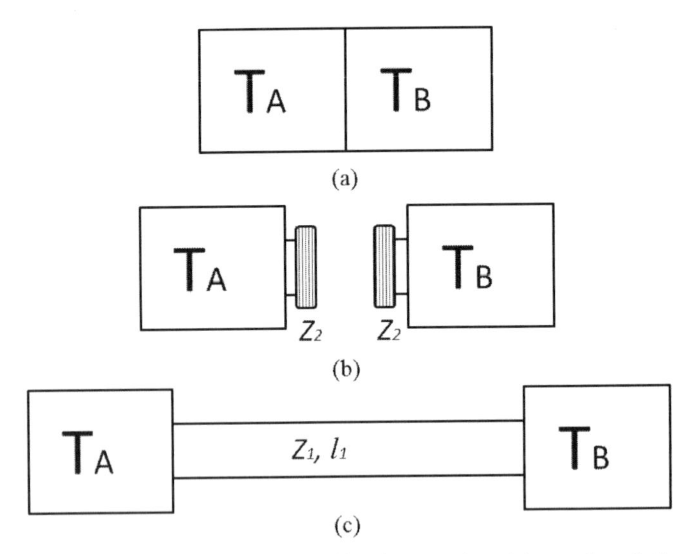

Figure 5.12 Structures used in the method implementation: (a) zero-length thru, (b)loads and (c) line. Ref [75]. © 2020 IEEE. Reprinted with permission.

Z_i; theses two matrices, given in eqn (5.57), may be expressed as

$$\mathbf{T_X} = \mathbf{T_A}\ \mathbf{T_{Zi}} = \begin{pmatrix} A_{Xi} & B_{Xi} \\ C_{Xi} & D_{Xi} \end{pmatrix} = D_X \begin{pmatrix} \widetilde{A}_{Xi} & \widetilde{B}_{Xi} \\ \widetilde{C}_{Xi} & 1 \end{pmatrix}, \tag{5.60}$$

$$\mathbf{T_{Yi}} = \mathbf{T_{Zi}^{-1}}\ \mathbf{T_B} = \begin{pmatrix} A_{Yi} & B_{Yi} \\ C_{Yi} & D_{Yi} \end{pmatrix} = D_{Yi} \begin{pmatrix} \widetilde{A}_{Yi} & \widetilde{B}_{Yi} \\ \widetilde{C}_{Yi} & 1 \end{pmatrix}. \tag{5.61}$$

It is worth noting that the calibration standard referred to as "Thru" by itself cannot provide information about the calibration reference impedance, but it can provide information about the calibration reference plane. However, when using the thru-line (or thru-load) combination, the reference impedance is imposed for the impedance of the line (or the load), respectively. Then, for an ideal zero-length thru (i=1), $\mathbf{T_{X1}}$; $\mathbf{T_{Y1}}$ becomes

$$\mathbf{T_{X1}} = \mathbf{T_A T_{Z1}}; \mathbf{T_{Y1}} = \mathbf{T_{Z1}^{-1} T_B}, \tag{5.62}$$

and for $i = 2$, $\mathbf{T_{X2}}$; $\mathbf{T_{Y2}}$, may be written as

$$\mathbf{T_{X2}} = \mathbf{T_A T_{Z2}}; \mathbf{T_{Y2}} = \mathbf{T_{Z2}^{-1} T_B}. \tag{5.63}$$

Due to the fact that $\mathbf{T_A}$ and $\mathbf{T_B}$ are unique and utilizing eqn (5.62) and (5.63) it results in

$$0.0\mathbf{T_{X2}^{-1} T_{X1}} = \mathbf{T_{Z2}^{-1} T_{Z1}}, \tag{5.64}$$

where the matrix product $\mathbf{T_{XP}} \triangleq \mathbf{T_{X2}^{-1} T_{X1}}$ may be expressed as

$$\mathbf{T_{XP}} = \frac{D}{\widetilde{C}_{X2}\rho} \begin{pmatrix} \widetilde{C}_{X1}\alpha & \beta \\ \widetilde{C}_{X2}\gamma & \widetilde{C}_{X2}\varphi \end{pmatrix}, \tag{5.65}$$

with α, β, γ, φ, defined as

$$\begin{aligned} \alpha &\triangleq \frac{\widetilde{A}_{X1}}{\widetilde{C}_{X1}} - \widetilde{B}_{X2}; \beta \triangleq \widetilde{B}_{X1} - \widetilde{B}_{X2}; \gamma \triangleq \frac{\widetilde{A}_{X2}}{\widetilde{C}_{X2}} - \frac{\widetilde{A}_{X1}}{\widetilde{C}_{X1}}; \\ \varphi &\triangleq \frac{\widetilde{A}_{X2}}{\widetilde{C}_{X2}} - \widetilde{B}_{X1}; \rho \triangleq \frac{\widetilde{A}_{X2}}{\widetilde{C}_{X2}} - \widetilde{B}_{X2}; D \triangleq \frac{D_{X1}}{D_{X2}}. \end{aligned} \tag{5.66}$$

Furthermore, the matrix product $\mathbf{T_{ZP}} \triangleq \mathbf{T_{X2}^{-1} T_{X1}}$, using eqn (5.59), may be written as

$$\mathbf{T_{ZP}} = \frac{1}{2} \begin{bmatrix} 1+R & 1-R \\ 1-R & 1+R \end{bmatrix}; R = \frac{Z_1}{Z_2}. \tag{5.67}$$

Comparing each term of matrices on both sides of eqn (5.64), Pulido-Gaytan et al [67] proposed a method to determine either Z_1 or Z_2 (line or load impedance, respectively). The method published in [67] requires the knowledge of $\widetilde{A_{X1}}$, $\widetilde{B_{X1}}$, $\widetilde{C_{X1}}$, and $\widetilde{A_{X2}}$, $\widetilde{B_{X2}}$, $\widetilde{C_{X2}}$, which are determined using two calibration techniques (e.g., TRL-TRM and TRL-TRRM). A procedure to determine the line characteristic impedance without using calibration comparison is given next. Given that $\mathbf{T_{XP}} = \mathbf{T_{ZP}}$, the eigenvalues of $\mathbf{T_{ZP}}$ and $\mathbf{T_{XP}}$ are expected to be identical and they are calculated from

$$\det(\mathbf{T_{ZP}} - \lambda_{\mathbf{Z}}\mathbf{I}) = 0; \det(\mathbf{T_{ZP}} - \lambda_{\mathbf{X}}\mathbf{I}) = 0. \tag{5.68}$$

The eigenvalues of $\mathbf{T_{ZP}}$ referred to as λ_Z^+, λ_Z^- are expressed as

$$\lambda_Z^+ = 1; \lambda_Z^- = R \tag{5.69}$$

and the eigenvalues of $\mathbf{T_{XP}}$ denoted as λ_X^+, λ_X^- are determined from

$$\lambda_X^{\pm} = \frac{\frac{D}{\rho}[C\alpha + \varphi] \pm \sqrt{(C\alpha + \varphi)^2 - 4(C\alpha\varphi - C\gamma\beta)}}{2}. \tag{5.70}$$

The term $\frac{\widetilde{C_{X1}}}{\widetilde{C_{X2}}} = C$ that appears in eqn (5.70) is derived from eqn (5.64). In their calculation, the condition $\mathbf{T_{XP}} = \mathbf{T_{ZP}}$ is used and given that the diagonal elements of $\mathbf{T_{ZP}}$ are equal, it results in $\widetilde{C_{X1}\alpha} = \widetilde{C_{X1}\phi}$; therefore, $C = \frac{\widetilde{C_{X1}}}{\widetilde{C_{X2}}}$ can be written as

$$C = \frac{\widetilde{C_{X1}}}{\widetilde{C_{X2}}} = \frac{\varphi}{\alpha}. \tag{5.71}$$

Once C is determined from eqn (5.71), it may be used in eqn (5.70) to calculate the eigenvalues $\lambda_X^{\pm}$ as follows:

$$\lambda_X^{\pm} = \frac{D}{\rho}\varphi(1 \pm \sqrt{\frac{\gamma\beta}{\varphi\alpha}}), \tag{5.72}$$

where λ_X^+ and λ_X^- are obtained using the plus (+) and minus (−) signs, respectively, of the square root in eqn (5.72). As already mentioned, $\mathbf{T_{XP}} = \mathbf{T_{ZP}}$ and consequently they have equal eigenvalues $\lambda_X^+ = \lambda_Z^+$ and $\lambda_X^- = \lambda_Z^-$. These two conditions allow us to determine D and then R

$$R = \frac{Z_1}{Z_2} = \frac{1 - \Gamma}{1 + \Gamma}; \Gamma = \frac{1 - R}{1 + R}, \tag{5.73}$$

where Γ is defined as

$$\Gamma \triangleq \sqrt{\frac{\gamma\beta}{\varphi\alpha}} = \sqrt{\frac{(\frac{\widetilde{A}_{X2}}{\widetilde{C}_{X2}} - \frac{\widetilde{A}_{X1}}{\widetilde{C}_{X1}}) - (\widetilde{B}_{X1} - \widetilde{B}_{X2})}{(\frac{\widetilde{A}_{X2}}{\widetilde{C}_{X2}} - \widetilde{B}_{X1})(\frac{\widetilde{A}_{X1}}{\widetilde{C}_{X1}} - \widetilde{B}_{X2})}}. \tag{5.74}$$

Using eqn (5.73) and the square of Γ, the quadratic expression can be derived from R

$$R^2 + \frac{2R(1+\Gamma^2)}{1-\Gamma^2} + 1. \tag{5.75}$$

The quadratic equation (5.75) has two solutions referred to as R^+ and R^-. Depending on the measurement frequency range and the length of the line, R^+ or R^- can exhibit discontinuities since these solutions depend on Γ, which also depends on $\frac{\widetilde{A}_X}{\widetilde{C}_X}$, $\widetilde{B}_X$ (they are computed using the measurements of the "thru-line" combination). But it is well known that these elements are ill-conditioned when the electrical length of the line is around $0°$ and multiples of $\pi(k\pi; k = 1, 2, n)$ [21], [71], [72].

For this reason, the solution to eqn (5.75) is obtained by calculating $\frac{\widetilde{A}_X}{\widetilde{C}_X}$, $\widetilde{B}_X$ according to the procedure reported in [21] and [74] and using R^+ or R^-. The correct root of eqn (5.75) may be chosen based on the expected behavior of the imaginary part of the line impedance (i.e., mostly inductive low frequencies and becomes capacitive as the frequencies increase). In this case, R^+ is selected for frequencies where the electrical length of the line is between $160°$ and $340°$.

It should be noted that Γ, given by eqn (5.74), depends only on $\frac{\widetilde{A}_{X1}}{\widetilde{C}_{X1}}$, $\frac{\widetilde{A}_{X2}}{\widetilde{C}_{X2}}$, $\widetilde{B}_{X1}$, $\widetilde{B}_{X2}$. This corresponds to an improvement to the method reported in [67], where the equivalent expressions required the full knowledge of the seven-error terms of the calibration given by eqn (5.60) and (5.61). It results from eqn (5.73) that the calculation of Z_1 (line impedance) needs the previous knowledge of Z_2 (load impedance) along with $\frac{\widetilde{A}_{X1}}{\widetilde{C}_{X1}}$, $\frac{\widetilde{A}_{X2}}{\widetilde{C}_{X2}}$, $\widetilde{B}_{X1}$, $\widetilde{B}_{X2}$. These terms are determined using the raw S-parameter measurements of thru-line [72] or thru-load [65] combinations, respectively.

For implementing this method, a long air line (85052–60036) and broadband loads, included in the 85052C 3.5-mm calibration kit from Agilent [84], were used. The S-parameters of a thru (Figure 5.12(a), a pair of loads of known impedance (Figure 5.12(b), and a line (Figure 5.12(c) were measured

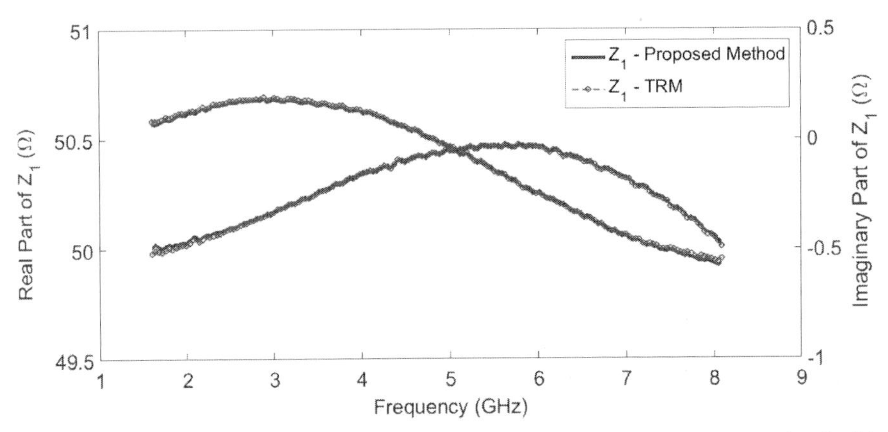

Figure 5.13 (a) Real and imaginary parts of Z_i. Ref [75]. © 2020 IEEE. Reprinted with permission.

with an uncalibrated vector network analyzer (PNA-X 5245AS) in the frequency range of the line of 1.6–8 GHz, which was determined using the information of the thru-line combination electrical length. The terms required by the proposed method are determined by using the measurements of thru-line and thru-load pair combinations. For determining the line characteristic impedance, the value of the load impedance used as a reference was assumed to be 50 Ω, a valid assumption when using coaxial broadband terminations [84]. The TRM calibration method [65] has been used to validate the proposed method.

The real and imaginary parts of the line impedance, computed by using the proposed method and the TRM calibration method, are shown in Figure 5.13. An excellent correlation is observed between these two methods. The absolute error of the real and imaginary parts of Z_L, using TRM as a reference, shown in Figure 5.14, corroborates the excellent correlation between these two methods, where it is noticeable that the maximum difference between the methods is very small, i.e., less than 0.03 Ω. Furthermore, in the frequency range where the line is recommended (2–7 GHz) [84], the absolute error is less than 0.01.

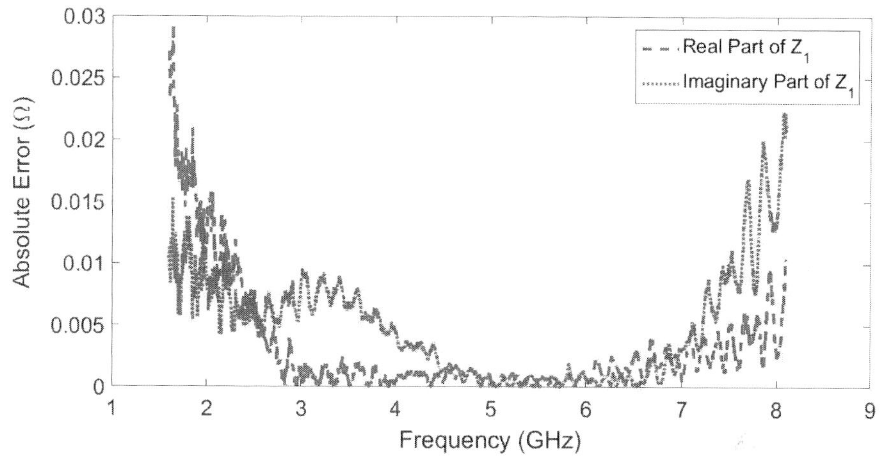

Figure 5.14 (b) Error in Z_i calculated as ABS $(Z_{1\mathrm{TRM}} - Z_1)$. Ref [75]. © 2020 IEEE. Reprinted with permission.

6

Load-pull Characterization of Power Transistors

Microwave solid-state devices are about to enter their half a century of age. Similarly to what was the common procedure in the early days of microwave solid-state PAs, power amplifier design is still focused on observing the load-pull (LP) contours of output power, efficiency, linearity, or any other metric, to then select a load impedance that maximizes some compromise between them [59]. The work reported in [13] is intended to help in this LP-based PA design process. There, Cripps proposed an intuitive and simple explanation for the dependence of delivered output power of a device on its terminating load. Despite its simplicity, this model has survived for decades of research. It has proved to be capable of estimating the PA output power LP contours. However, engineers and researchers still rely on LP measurements and simulations to find the device's behavior as a function of its loading conditions.

While nowadays significant effort is devoted toward the development of high-performance microwave transistors to improve PA efficiency and output power [9], accurate characterization tools are necessary to evaluate the device's performance in the nonlinear regime [35]. Large-signal characterization of microwave transistors is essential for determining the device's performance in non linear applications, considering the limitations of S-parameter in such applications [33, 32]. LP is presently the most commonly used technique for carrying out this task.

For a PA design, the optimal loading conditions of transistors primarily depend on the transistor's maximum current and voltage waveforms. This is significantly different from the linear case, where the optimal loading conditions (e.g., conjugate match, and noise match) are directly identified from S-parameters [36]. In the nonlinear case, LP systems determine the appropriate impedance values experimentally through the use of impedance tuning mechanisms, which vary the load reflection coefficient Γ_{LOAD} while

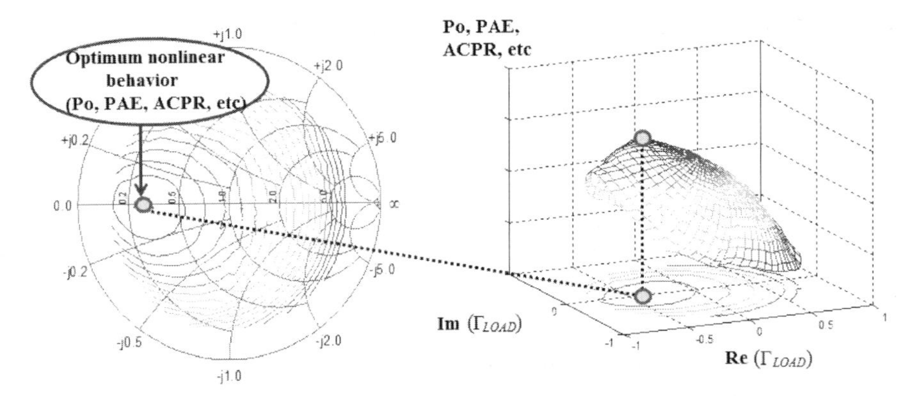

Figure 6.1 Load-pull contours as a method for reporting the results of an active device characterization using load-pull systems.

the behavior of the DUT is evaluated in terms of a determined large-signal metric.

The results of the LP characterization of a DUT are commonly reported as contours in the Smith chart. These contours correspond to the level curves of the surface formed by the evaluation of a given parameter versus the variation of the impedance at the device's output plane. Figure 6.1 illustrates this concept.

The load of impedance Z_{LOAD} at the output of a DUT, the incident and reflected waves a_2 and b_2 at the DUT's output port, and the reflection coefficient Γ_{LOAD} are related as

$$\Gamma_{\mathrm{LOAD}} = \frac{a_2}{b_2} = \frac{Z_{\mathrm{LOAD}} - Z_0}{Z_{\mathrm{LOAD}} + Z_0}. \tag{6.1}$$

As expressed in eqn (6.1), the reflection factor (and the impedance) at the output of a DUT may be varied by varying the ratio of the amplitudes and the difference of the phases of two waves a_2 and b_2. The wave b_2 emerging from the DUT toward the load is generated by the DUT itself; therefore, the problem to deal with in LP is to find a method by which the wave a_2 entering the DUT's output port is generated. In this respect, LP systems may be classified according to the method by which load impedances are synthesized (i.e., how the wave a_2 is generated). Load impedance synthesis may be achieved through the use of either passive or active elements [17, 57]. The following sections are devoted to discuss the operation along with the advantages and disadvantages of passive and active LP systems used in the characterization of microwave transistors.

6.1 Passive Load-pull Systems

The classical method to synthesize impedances in LP measurement systems is to use passive impedance tuners. Passive tuners, introduced for the first time in [15], are based on the mechanical movement of a probe along a transmission line, in order to generate complex reflection factors. The probe varies the reflection factor through its movement in the horizontal and vertical directions. The movement of a probe in the vertical direction alters mainly the magnitude of the reflection factor, while its movement in the horizontal direction alters mainly its phase. These impedance tuners are known as passive-mechanical. A generic representation of the working principle of a passive-mechanical tuner is depicted in Figure 6.2.

Passive LP systems are capable of working in very high-power environments. Nevertheless, due to inherent losses of the elements located between the DUT and tuners (e.g., bias-tees, cables, adapters, or test fixtures) along with the losses of the tuner itself, these systems are unable to synthesize high Γ_{LOAD} values at high frequencies. This problem is of paramount importance when characterizing power transistors, since these devices manage very high currents, and thus very low impedances, with a corresponding optimum Γ_{LOAD} located on or near the edge of the Smith chart [41]. This drawback is also a limiting factor when characterizing devices under harmonic terminations, where loads of very low/high impedance are required [2, 3, 26] at

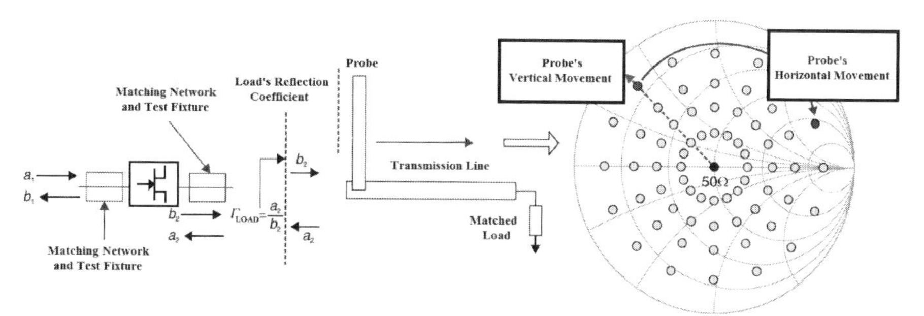

Figure 6.2 Principle of operation of a passive LP systems and the reflection coefficient synthesis. In this type of tuners, the vertical movements of a probe vary mostly the magnitude of Γ_{LOAD} and the horizontal movements vary mostly the phase of Γ_{LOAD}.

high frequencies. These limitations of passive LP systems may be overcome by using active tuning systems.

6.2 Active Load-pull Systems

Phase-shifting along with attenuation and amplification systems can be used to control the magnitude and phase of the wave a_2, and thus the magnitude and phase of the reflection coefficient at the output of a DUT. Impedance tuning systems based on this approach are known as active loads and the load-pull systems based on this method are referred to as active LP systems.

Although researchers have reported advances in the use of this LP approach for characterizing high-power transistors [2, 40], active LP systems are ideally suited for low- and medium-power applications. These systems are capable of synthesizing any Γ_{LOAD} at the DUT plane, even Γ_{LOAD} values greater than unity. Thus, when using this type of LP systems, there is an imminent risk of oscillation[1], Which can lead to damage of the device being tested.

There are different types of active LP techniques among which open-loop, closed-loop, and feedforward are the most commonly used.

6.2.1 Open-loop load-pull systems

The open-loop LP technique uses an external RF source, different from the source used to drive the DUT, for generating the wave a_2 at the output of the DUT. Both the source used to drive the DUT and the source in the open-loop at the DUT's output must be locked to a common reference signal in order to maintain phase coherence between the signal a_2 and the signal b_2.

Figure 6.3 illustrates the principle of operation of the open-loop LP approach. The amplifier in the open-loop is used to boost the signal emerging from the RF source. The attenuator and phase shifter are used to modify the phase and magnitude of the wave a_2. Isolators prevent any damage to the loop amplifier and the loop RF source. It is evident that the open-loop LP can synthesize reflection coefficients Γ_{LOAD} of any magnitude, just by changing the magnitude of the wave a_2. The magnitude of Γ_{LOAD} will be zero when the source in the loop is turned off, or close to zero if the attenuation in the loop is much higher than the gain of the loop amplifier.

[1]Although LP systems are mainly used in the characterization of power transistors for the design of PAs, they can also be used to design oscillators or mixers [76, 91]. In those cases, Γ_{LOAD} values greater than unity become of interest.

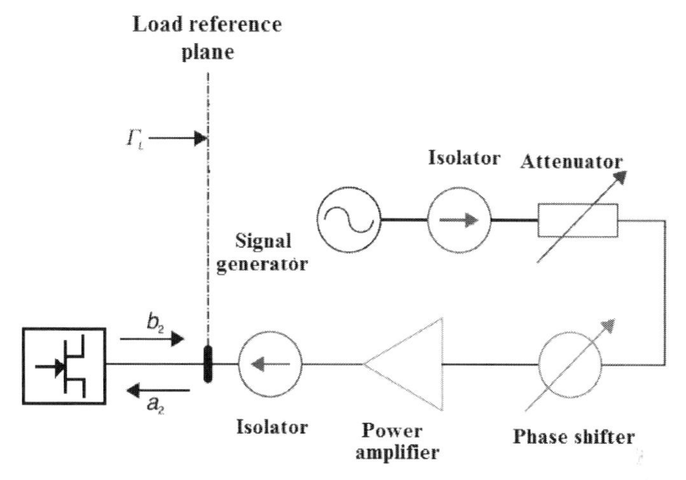

Figure 6.3 Generic diagram of an active LP system using the open-loop active LP concept.

The magnitude of Γ_{LOAD} will increase as the signal emerging from the active loop increases. The magnitude and phase of the signal b_2 emerging from the DUT change as the DUT's drive, bias, and loading conditions are changed. Thus, synthesizing a determined reflection factor at the DUT output port may become an iterative (and sometimes slow) process, which may require the use of complex algorithms [81].

Two drawbacks when using the open-loop technique are related to the harmonic and power scaling of the active loop. When characterizing devices under harmonic terminations, a number of sources identical to the number of harmonics must be used, all of them locked to a common reference signal. Frequency multipliers can be used to alleviate this limitation by allowing the generation of waves at different commensurate frequencies. Regarding the power, when characterizing high-power devices under high Γ_{LOAD} conditions, it may be difficult to generate in the active loop a wave of magnitude comparable to the wave generated by the DUT [41, 2]; it requires the use of loop amplifiers managing power levels of the order of the power level produced by the DUT.

6.2.2 Feedforward load-pull systems

The original idea of the feedforward LP approach was proposed in [83]. The operating principle of the feedforward LP approach is depicted in Figure 6.4. The signal generated by an RF source (and boosted by an amplifier if

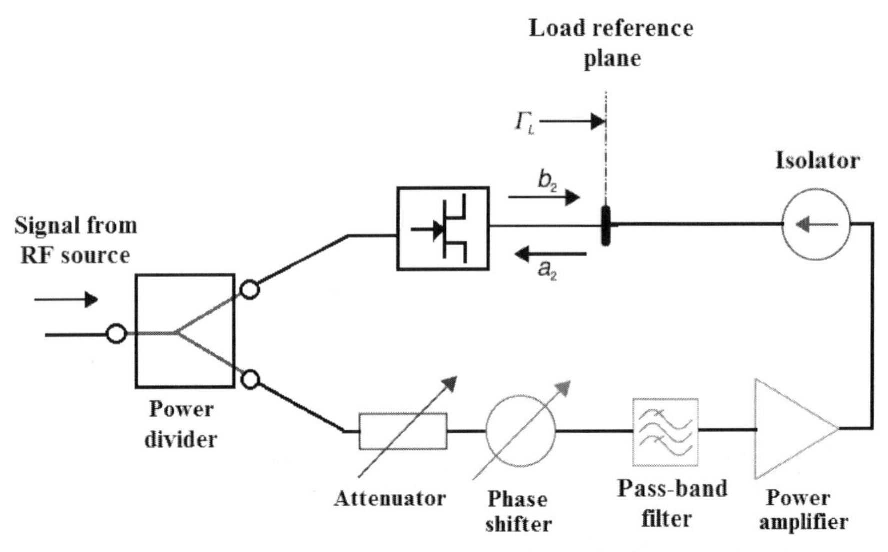

Figure 6.4 Generic diagram of an active LP system using the feedforward active LP concept.

required) at the input of the DUT is split into two different paths. One path drives the input port of the DUT, while in the other path the signal is varied in magnitude and phase, to be further injected into the DUT's output port as a signal a_2. Once combined with the signal arising from the DUT b_2, the equivalent reflection coefficient Γ_{LOAD} results.

The operation principle of feedforward LP systems is similar to the open-loop technique. The main difference relies on the origin of the signal a_2. Since the feedforward LP approach is based on the use of a second path for the signal entering the DUT's output port, this LP approach is also known [56] as the *two signal path* LP technique.

6.2.3 Closed-loop load-pull systems

The synthesis of reflection coefficients at the output of the DUT may be achieved by sampling the wave generated by the transistor traveling toward the load, controlling its magnitude and phase, and re-injecting it toward the transistor. This approach, known as the closed-loop LP technique [6], is depicted in Figure 6.5.

Let us outline the realization of a closed-loop active LP system. As may be observed in Figure 6.5, for the realization of a closed-loop LP system, it is necessary to use a three-port device (either a circulator or a directional

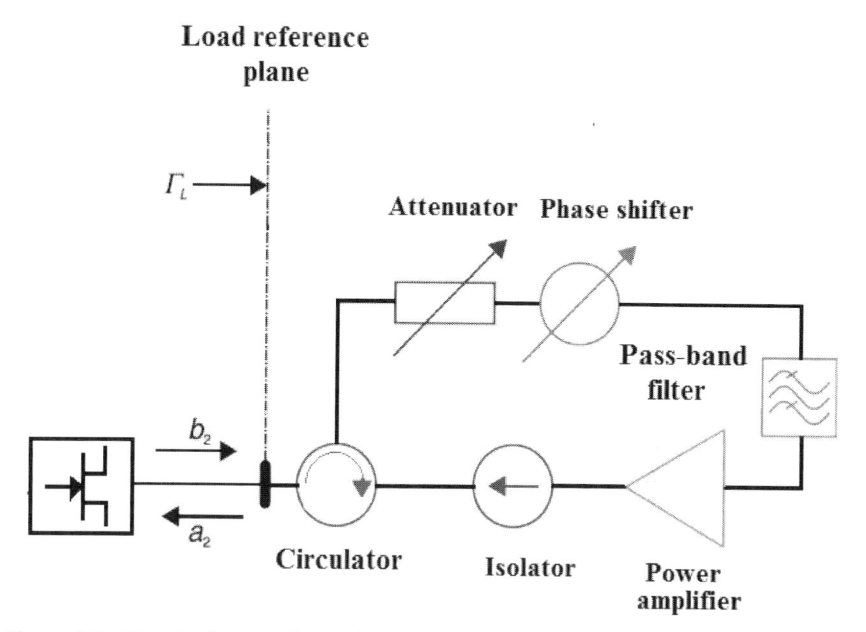

Figure 6.5 Generic diagram of an active LP system using the closed-loop active LP concept.

coupler) in the active loop to route the waves b_2 and a_2 in their respective directions. As in the open-loop and feedforward LP systems, the magnitude of the wave injected to the DUT's output is varied through the use of an attenuator and an amplifier located in the loop. Alternatively, a *passive closed-loop* may be used to vary the impedance at the DUT's output by using only the attenuator and phase shifter [34] in the loop. Depending on the frequency and the attenuator/phase shifter insertion loss, this passive loop may allow synthesizing loads located close to the edge of the Smith chart.

Unlike the open-loop technique, the closed-loop technique does not require phase-locking to maintain phase coherence between the signal a_2 and the signal b_2; the waves a_2 and b_2 are coherent by nature. Closed-loop is more similar to the feedforward approach; indeed, in some textbooks [33]. The closed-loop technique is referred to as the *feedback LP technique*. The main difference between these two LP techniques relies on the point in which the signal injected to the DUT output is sampled at: the input (feedforward) or the output (closed-loop) of the DUT.

Furthermore, since in the closed-loop technique the signal injected at the DUT output is a modified version of the DUT output signal, it is ensured that the synthesized load impedance is independent of drive level. It avoids the iterative process required in open-loop and feedforward techniques to generate a determined load impedance [95].

The appropriate LP system to be used depends on the type of measurement to be performed. Passive LP systems are mostly used in very high-power applications; meanwhile, active LP systems are preferred when low- and medium-power devices are characterized under high Γ_{LOAD} loading conditions. A special type of LP systems in which the passive and active approaches are combined (hybrid LP systems) have emerged in recent years [41]. They are used to alleviate the power requirements from active elements of active LP systems in those cases in which high Γ_{LOAD} values have to be synthesized at the output of high-power devices [26].

6.3 Power-based Calibrated vs Vector-Receiver Load-pull Systems

In the preceding section, a classification of LP systems in terms of the procedures by which the load at the output of a DUT is varied was presented. LP systems are also classified as a function of the form by which the measurement and calibration are performed. In this respect, LP measurement systems are classified as power-based LP systems and vector-receiver LP systems.

In power-based LP systems (Figure 6.6), the impedances presented at the output of the DUT during the LP characterization have to be known prior to the measurement process. In this order, pre-calibrated passive impedance tuners must be used; it implies using a VNA to determine the S-parameters of the tuner at a set of positions of the probe along the vertical and horizontal axes (see Figure 6.2).

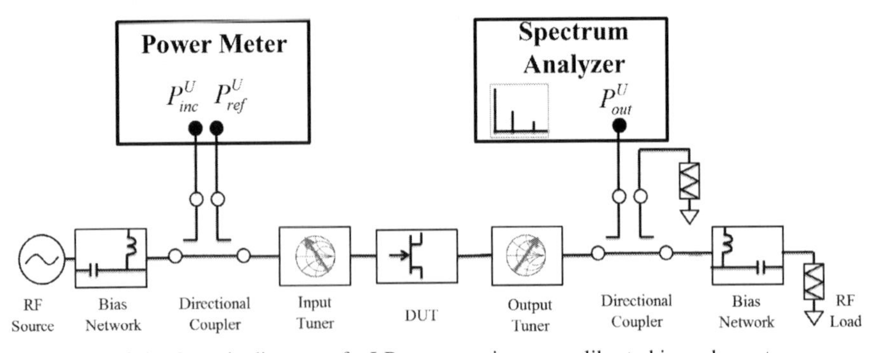

Figure 6.6 Generic diagram of a LP system using pre-calibrated impedance tuners.

Tuner repeatability is a must in this type of LP systems since it is required that the passive-tuner behaves identically when measured by a VNA and when placed in the LP setup. In addition, since every element located between the tuner and the DUT and between the tuner and the load (e.g., cables, connectors, adapters, test-fixtures, or bias-tees) affects the impedance seen by the DUT, all of the elements of the LP system must be also pre-characterized.

Power-based LP systems allow measuring only scalar parameters of a DUT, such as output power, power gain, and efficiency[2] while varying the impedance at its output. In order to measure the power levels at the input and output of a DUT, power receivers (power meters and/or spectrum analyzers) are used, as shown in Figure 6.6. Since this type of LP systems only allow the use of passive tuning systems, tuners are placed as close as possible to the DUT, in order to achieve the highest reflection coefficient at the DUT's input and output ports. Hence, the power levels measured by the power receivers located at the input ($P^U_{\mathrm{IN}} = P^U_{\mathrm{inc}} - P^U_{\mathrm{ref}}$) and output ($P^U_{\mathrm{OUT}}$) sides of a DUT are not the actual power levels at the DUT plane but their uncalibrated quantity. In order to determine the power at the DUT ports, the knowledge of the behavior of the passive elements placed between the DUT and the power receivers is used to accurately de-embed their effects [54].

The limitations of power-based LP systems regarding tuner repeatability, pre-characterization of each element of the system, the inability to measure vectorial figures of merit of a DUT, along with the inability to use active load tuning systems are overcome by using vector-receiver LP systems, which measure power and other relevant quantities in real time.

Vector-receiver load-pull (VRLP) systems [55] comprise a signal source (and a boost amplifier if required) to drive the DUT, impedance tuners, directional couplers, and a vector-receiver. The vector-receiver is used to capture uncalibrated ratios of parameters of interest along with the magnitude of the traveling waves at the input and output of a DUT. This information is then used to determine the impedances and the behavior at the DUT reference plane by using a calibration procedure. Nowadays, several VNAs [46, 5] provide external access to the internal instrument receivers, which are in turn used as vector-receiver in LP measurements.

In VRLP systems, the impedances presented at the DUT's output are determined at the same time the load impedance is varied. In this type of systems, neither the impedance tuner nor the elements of the LP system are

[2]Vectorial figures of merit such as amplitude-to-phase conversion of a DUT (AM-PM distortion) or the variation of the DUT's input impedance as a function of the input power cannot be measured.

required to be characterized prior to the measurement process. Therefore, tuner repeatability is not a major concern in VRLP systems.

The accuracy of VRLP measurement systems relies on the calibration of the measurement setup. In order to accurately characterize a DUT, a calibration process is used [28, 61, 73]. Once calibrated, VRLP systems allow measuring the impedance at the input of the DUT, the load impedance at the DUT's output, and the large-signal behavior of the DUT. Moreover, unlike power-based LP systems in which only passive impedance tuners may be used, VRLP systems allow the use of both passive and active tuning systems. This feature of VRLP systems is very useful to characterize power transistors, requiring very low impedances for optimum operation.

Although both power-based and VRLP systems are widely used in the characterization of power transistors, the measurement accuracy is enhanced when using VRLP systems [25, 55], provided that they are accurately calibrated.

6.4 Calibration of a VRLP System

Nonlinear devices' characterization requires complex measurement systems, where the most utilized are the load-pull systems. Meanwhile, in power-based load-pull systems, measurement accuracy relies on pre-calibrated impedance tuners and power receivers, and in VRLP systems (Figure 6.8), the impedances presented to the device under test (DUT) and the power at its ports are measured in real time using a vector-receiver.

Characterizing devices using a VRLP system require removal of the effects of the system's elements using calibration methods [28, 76, 61, 66].

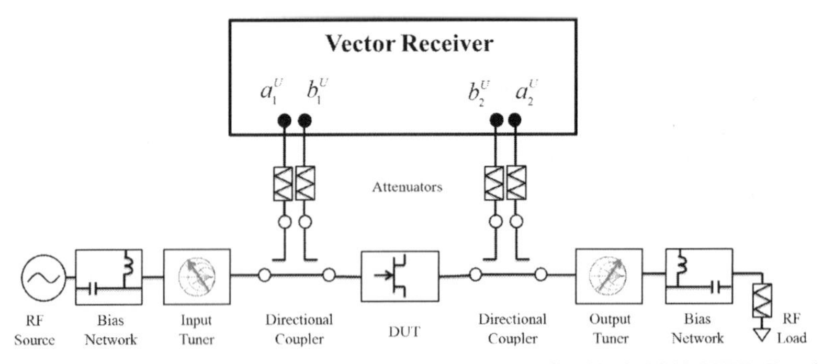

Figure 6.7 Generic diagram of a real-time LP system. Ref [73]. © 2022 IEEE. Reprinted with permission.

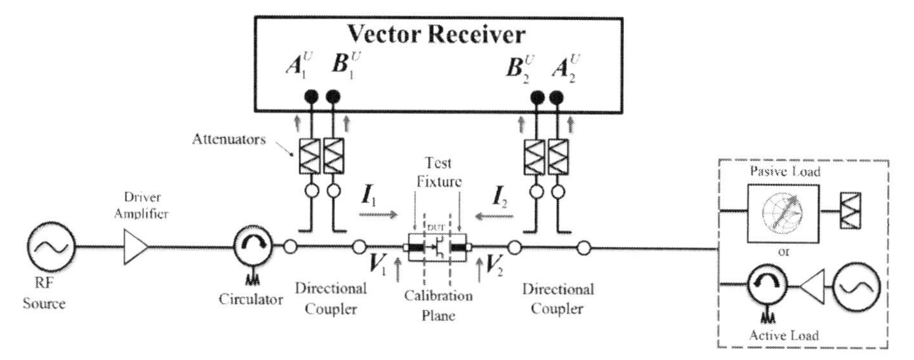

Figure 6.8 Vector-receiver load-pull system. Red dotted lines denote the calibration plane.

The calibration of a VRLP system comprises two parts: relative calibration, used to determine ratios of parameters, and power calibration, used to determine the power levels at the DUT ports. Relative calibration may be carried out using two-port calibration techniques, whereas power calibration is performed using a power reference device (e.g., power meter) connected at a coaxial plane located close to the calibration plane. In this section, a complete mathematical formulation of the calibration procedure, developed using the ABCD-parameters matrix formalism, is presented.

6.4.1 Relative calibration

At present, vector network analyzers allowing access to the instrument receivers are used as the vector-receiver (VR) in load-pull systems (Figure 6.9(a)). Thus, VNA calibration techniques, such as the thru-reflect-line (TRL) [21, 72, 71], thru-reflect-match (TRM) [22, 23, 65], or short-open-load-thru (SOLT) [27] may be used to carry out the relative calibration of VRLP systems.

Relative calibration relates ratioed quantities measured at the VR (U_1–U_2 planes), to ratioed quantities at the calibration plane (D_1–D_2 planes), as depicted in Figure 6.9. The voltage and current quantities at the VR ports, V_U^k, and I_U^k, $k = 1, 2$, are defined as a function of the measured traveling waves, A_U^k, and B_U^k, as follows [79]:

$$V_k^U = \sqrt{Z_0}\left(1 + \frac{B_k^U}{A_k^U}\right) A_k^U = \sqrt{Z_0}\left(1 + \frac{A_k^U}{B_k^U}\right) B_k^U, \qquad (6.2)$$

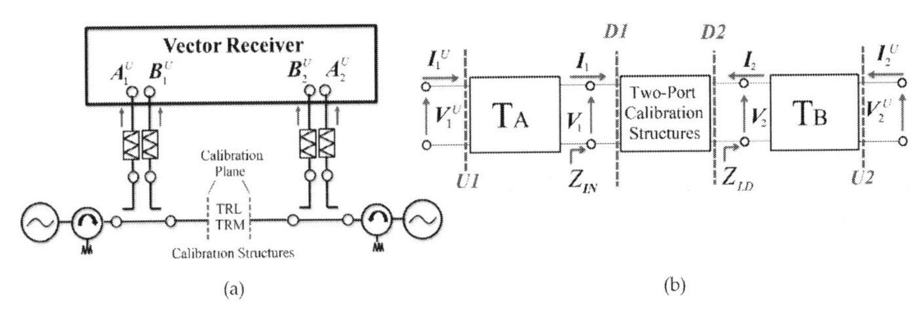

(a) (b)

Figure 6.9 (a) System configuration for relative calibration; (b) measurement of calibration structures. Red dotted lines denote the calibration plane; blue dotted lines denote the vector-receiver plane. Ref [73]. © 2022 IEEE. Reprinted with permission.

$$I_k^U = \frac{1}{\sqrt{Z_0}}\left(1 - \frac{B_k^U}{A_k^U}\right) A_k^U = \sqrt{Z_0}\left(1 - \frac{A_k^U}{B_k^U}\right) B_k^U, \tag{6.3}$$

where Z_0 represents the measurement system impedance. Using the eight-term error model, V_U^k and I_U^k are defined as a function of the voltage and current quantities at the calibration reference plane, V_k and I_k, as follows:

$$\begin{bmatrix} V_1 \\ I_1 \end{bmatrix} = \mathbf{T_A}^{-1} \begin{bmatrix} V_1^U \\ I_1^U \end{bmatrix}, \tag{6.4}$$

$$\begin{bmatrix} V_2 \\ I_2 \end{bmatrix} = \mathbf{T_B} \begin{bmatrix} V_2^U \\ I_2^U \end{bmatrix}. \tag{6.5}$$

In eqn (6.4) and (6.5), $\mathbf{T_A}$ and $\mathbf{T_B}$ are matrices representing the networks connecting the DUT ports to the VR ports, as shown in Figure 6.10. These matrices may be expressed as

$$\mathbf{T_A} = \mathbf{T_X}\mathbf{T_Z}^{-1} = D_X \begin{bmatrix} \overline{A_X} & \overline{B_X} \\ \overline{C_X} & 1 \end{bmatrix} \begin{bmatrix} -Z_B & Z_A \\ 1 & 1 \end{bmatrix}^{-1}, \tag{6.6}$$

$$\mathbf{T_B} = \mathbf{T_Z}\mathbf{T_Y} = D_Y \begin{bmatrix} -Z_B & Z_A \\ 1 & 1 \end{bmatrix} \begin{bmatrix} \overline{A_Y} & \overline{B_Y} \\ \overline{C_Y} & 1 \end{bmatrix}, \tag{6.7}$$

where the terms Z_A and Z_B in the matrix $\mathbf{T_Z}$ represent the impedance of one of the calibration structures used in a determined calibration technique. For the sake of clarity, the case in which $Z_A = Z_B = Z$ is considered first. The seven terms $\overline{A_X}, \overline{B_X}, \overline{C_X}, \overline{A_Y}, \overline{B_Y}, \overline{C_Y}$, and $D_X D_Y$ are determined by using only information provided by the measurement of the calibration structures

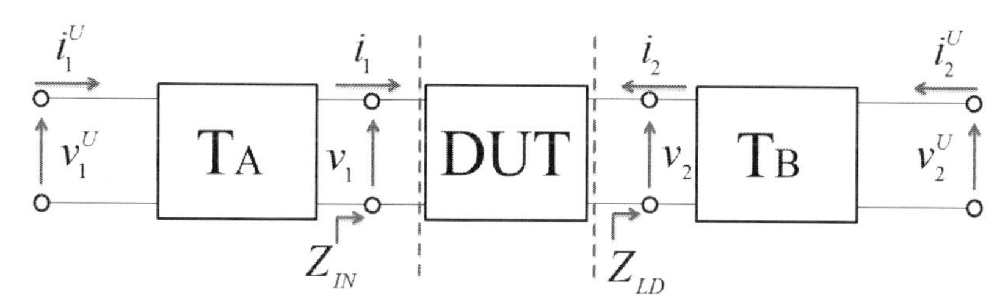

Figure 6.10 Model for the measurement setup configuration for the relative calibration. Ref [73]. © 2022 IEEE. Reprinted with permission.

corresponding to a calibration technique. Then, by substituting eqn (6.6) in eqn (6.4) and eqn (6.7) in eqn (6.5), and developing the resultant expressions using $Z_A = Z_B = Z$, the following equations may be derived:

$$V_1 = \frac{-Z}{D_X \cdot \Delta_X}\left[(1+\overline{C_X})V_1^U - (\overline{B_X}+\overline{A_X})I_1^U\right], \tag{6.8}$$

$$I_1 = \frac{1}{D_X \cdot \Delta_X}\left[(1-\overline{C_X})V_1^U - (\overline{B_X}-\overline{A_X})I_1^U\right], \tag{6.9}$$

$$V_2 = Z \cdot D_Y\left[(\overline{C_Y}-\overline{A_Y})V_2^U - (1-\overline{B_Y})I_2^U\right], \tag{6.10}$$

$$I_2 = -D_Y\left[(\overline{C_Y}+\overline{A_Y})V_2^U - (1+\overline{B_Y})I_2^U\right], \tag{6.11}$$

where $\Delta_X = \overline{A_X} - \overline{B_X C_X}$. Then, from eqn (6.8)–(6.11), expressions for the impedance at the input of the DUT and the load impedance at the DUT plane, defined as $Z_{\text{IN}} = V_1/I_1$ and $Z_{\text{LD}} = -V_2/I_2$, may be obtained

$$Z_{\text{IN}} = -Z \cdot \frac{(1+\overline{C_X})Z_1^U - (\overline{B_X}+\overline{A_X})}{(1-\overline{C_X})Z_1^U - (\overline{B_X}-\overline{A_X})}, \tag{6.12}$$

$$Z_{\text{LD}} = Z \cdot \frac{(\overline{C_Y}-\overline{A_Y})Z_2^U - (1-\overline{B_Y})}{(\overline{C_Y}+\overline{A_Y})Z_2^U - (1+\overline{B_Y})}, \tag{6.13}$$

where $Z_k^U = V_k^U/I_k^U$, $k = 1, 2$. Meanwhile, the DUT's voltage gain and current gain, defined as $G_V = V_2/V_1$ and $G_I = -I_2/I_1$, may be expressed as

$$G_V = -\Delta_X \cdot D_X D_Y \cdot \frac{(\overline{C_Y}-\overline{A_Y}) - (1-\overline{B_Y})Y_2^U}{(1+\overline{C_X}) - (\overline{B_X}+\overline{A_X})Y_1^U}, \tag{6.14}$$

$$G_I = -\Delta_X \cdot D_X D_Y \cdot \frac{(\overline{C_Y}+\overline{A_Y})Z_2^U - (1+\overline{B_Y})}{(1-\overline{C_X})Z_1^U - (\overline{B_X}-\overline{A_X})}, \tag{6.15}$$

where $Y_k^U = 1/Z_k^U$, $G_V^U = V_2^U/v_1^U$, and $G_I^U = -I_2^U/I_1^U$.

The commonly used definition for gain in terms of the transmitted and incident waves $G_D = B_2/A_1$ can be determined as

$$G_D = \frac{V_2 - Z_0 \cdot I_2}{V_1 + Z_0 \cdot I_1} = \frac{Z_{\text{LD}} + Z_0}{Z_{\text{IN}} + Z_0} G_I = \frac{1 + Z_0/Z_{\text{LD}}}{1 + Z_0/Z_{\text{IN}}} G_V. \tag{6.16}$$

It may be noted from eqn (6.12) and (6.13) that for determining Z_{IN} and Z_{LD}, the knowledge of Z is mandatory. Regarding the gain, according to eqn (6.14) and (6.15), the voltage gain and current gain do not depend on the knowledge of Z. On the other hand, the gain in terms of incident and transmitted waves G_D depends on Z through its own dependence on Z_{IN} and Z_{LD}. It is worth noting that the power gain G_P, defined as

$$G_p = |G_V|^2 \frac{\text{Re}(1/Z_{\text{LD}})}{\text{Re}(1/Z_{\text{IN}})} = |G_I|^2 \frac{\text{Re}(Z_{\text{LD}})}{\text{Re}(Z_{\text{IN}})}, \tag{6.17}$$

depends on Z through its own dependence on $\text{Re}(Z_{\text{IN}})$ and $\text{Re}(Z_{\text{LD}})$, unless the value of Z is purely real.

Relative calibration using non-symmetrical loads in the TRM technique:

In this section, a more general formulation to the relative calibration of an LP system, in which Z_A and Z_B are allowed to be either identical or different, is presented. So far, this condition has been encountered to occur solely when the relative calibration of an LP system is carried out using the TRM technique implemented using symmetrical or non-symmetrical loads of arbitrary impedance [66].

By substituting eqn (6.6) in eqn (6.4) and eqn (6.7) in eqn (6.5), and developing the resultant expressions, the following expressions for V_k and I_k, $k = 1, 2$, may be derived:

$$V_1 = \frac{-Z_A}{D_X \cdot \Delta_X}\left[\left(\frac{Z_B}{Z_A} + \overline{C_X}\right) V_1^U - \left(\frac{Z_B}{Z_A}\overline{B_X} + \overline{A_X}\right) I_1^U\right], \tag{6.18}$$

$$I_1 = \frac{1}{D_X \cdot \Delta_X}\left[(1 - \overline{C_X})V_1^U - (\overline{B_X} - \overline{A_X})I_1^U\right], \tag{6.19}$$

$$V_2 = Z_A \cdot D_Y \left[\left(\overline{C_Y} - \frac{Z_B}{Z_A}\overline{A_Y}\right) V_2^U - \left(1 - \frac{Z_B}{Z_A}\overline{B_Y}\right) I_2^U\right], \tag{6.20}$$

$$I_2 = -D_Y\left[(\overline{C_Y} + \overline{A_Y})V_2^U - (1 + \overline{B_Y})I_2^U\right]. \tag{6.21}$$

Note that the expressions for I_1 and I_2 are identical to the expressions previously presented in eqn (6.9) and (6.11). The expressions for V_1 and V_2, on the other hand, take into account possible asymmetry between the loads used as calibration elements in the TRM are identical to the expressions shown in eqn (6.8) and (6.10) only if $Z_A = Z_B$.

In this order, the impedance at the input of the DUT and the load impedance at the DUT plane may be expressed as

$$Z_{\text{IN}} = -Z_A \cdot \frac{\left(\frac{Z_B}{Z_A} + \overline{C_X}\right) Z_1^U - \left(\frac{Z_B}{Z_A}\overline{B_X} + \overline{A_X}\right)}{(1 - \overline{C_X})Z_1^U - (\overline{B_X} - \overline{A_X})}, \tag{6.22}$$

$$Z_{\text{LD}} = Z_A \cdot \frac{\left(\overline{C_Y} - \frac{Z_B}{Z_A}\overline{A_Y}\right) Z_2^U - \left(1 - \frac{Z_B}{Z_A}\overline{B_Y}\right)}{(\overline{C_Y} + \overline{A_Y})Z_2^U - (1 + \overline{B_Y})}. \tag{6.23}$$

Since the expressions for I_1 and I_2 are independent of the impedances Z_A and Z_B, the current gain G may be expressed as shown in eqn (6.15). Regarding the voltage gain G_V, it depends on the knowledge of the impedances Z_A and Z_B as follows:

$$G_v = -\Delta_X \cdot D_X D_Y \cdot \frac{\left(\overline{C_Y} - \frac{Z_B}{Z_A}\overline{A_Y}\right) - \left(1 - \frac{Z_B}{Z_A}\overline{B_Y}\right) Y_2^U}{\left(\frac{Z_B}{Z_A} + \overline{C_X}\right) - \left(\frac{Z_B}{Z_A}\overline{B_X} + \overline{A_X}\right) Y_1^U}. \tag{6.24}$$

It differs from the result previously reported in eqn (6.14), where G_v was found to be independent of the knowledge of the impedance $Z = Z_A = Z_B$.

Regarding the gain expressed as the ratio of the transmitted to incident waves G_d. It depends on the knowledge of the impedances Z_A and Z_B through the dependence of G_v, Z_{IN}, and Z_{LD} on Z_A and Z_B.

6.4.2 Power calibration

Thus far, expressions to determine the impedance at the input of a DUT, its gain, and the load impedance at the DUT plane have been derived. Nonetheless, in LP, an important concern is to specify the power levels, at the input and output of the DUT, at which these ratioed quantities are defined.

In this order, by analyzing the structure shown in Figure 6.10, the power at the input and output of a DUT may be expressed as

$$P_{\text{IN}} = |V_1|^2 \,\text{Re}(1/Z_{\text{IN}}), \tag{6.25}$$

$$P_{\mathrm{OUT}} = |V_2|^2\,\mathrm{Re}(1/Z_{\mathrm{LD}}), \tag{6.26}$$

where

$$|V_1| = \frac{|Z|}{|D_X|}\left|(1+\overline{C_X}) - (\overline{B_X}+\overline{A_X})Y_1^U\right|\frac{|V_1^U|}{|\Delta_X|}, \tag{6.27}$$

$$|V_2| = |Z|\,|D_Y|\left|(\overline{C_Y}+\overline{A_Y}) - (1-\overline{B_Y})Y_2^U\right|\left|V_2^U\right|. \tag{6.28}$$

From eqn (6.25)–(6.28), it is noted that for determining P_{IN} and P_{OUT}, it is mandatory to know $|D_X|$ and $|D_Y|$. These terms may be calculated by using the measurement setup configuration shown in Figure 6.11, as described next.

Calculation of $|\mathbf{D_X}|$ and $|\mathbf{D_Y}|$:

The procedure for determining $|D_X|$ and $|D_Y|$ consists of the following: (1) to make a thru connection at the calibration plane, (2) to connect three loads of known impedance at a coaxial plane, and (3) to connect a PM at that coaxial plane (Figure 6.12) [28].

When characterizing transistors, the calibration plane is typically set either at the probe-tips when dealing with on-wafer devices or at the center of a test fixture when dealing with packaged devices. As shown in Figure 6.12, there is a two-port network connecting the calibration plane to a coaxial plane. This two-port network may include, among other elements, the test fixture, directional coupler, connectors, and cables. This network may be represented by an ABCD-parameters matrix T_C.

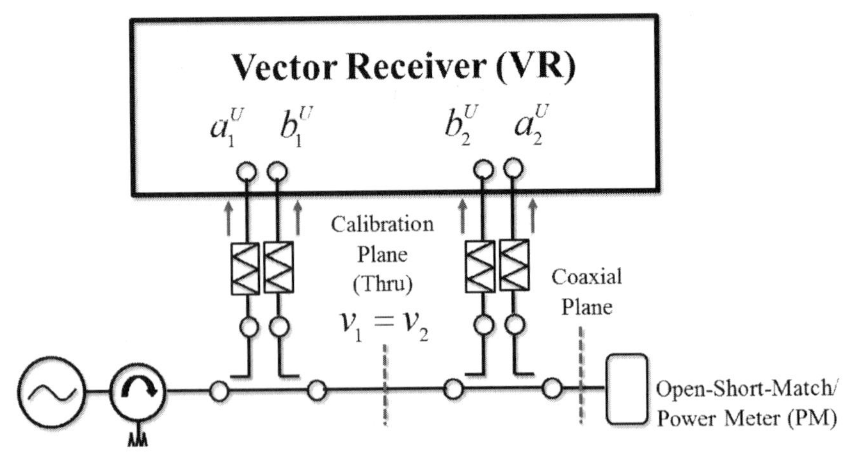

Figure 6.11 Measurement setup configuration for power calibration. Ref [73]. © 2022 IEEE. Reprinted with permission.

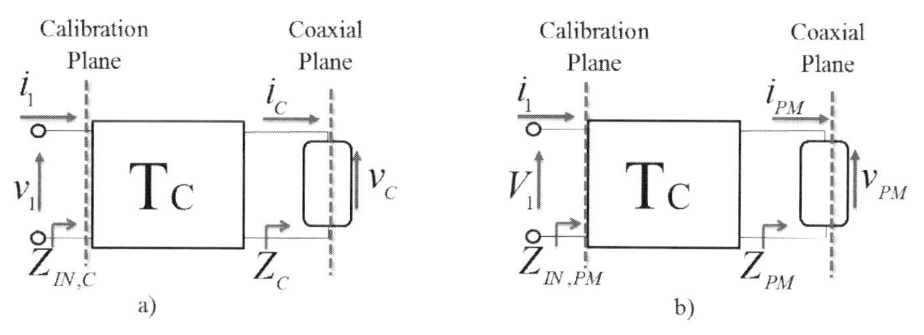

Figure 6.12 Model for the measurement of the (a) coaxial loads and (b) the power meter.

According to the structure shown in Figure 6.12(a), the voltages and currents at the calibration plane, V_1 and I_1, may be related to the voltages and currents at the coaxial plane, V_C and I_C, as

$$\begin{bmatrix} V_1 \\ I_1 \end{bmatrix} = \mathbf{T_C} \begin{bmatrix} V_C \\ I_C \end{bmatrix}, \tag{6.29}$$

with $\mathbf{T_C}$ defined as

$$\mathbf{T_C} = D_C \begin{bmatrix} \overline{A_C} & \overline{B_C} \\ \overline{C_C} & 1 \end{bmatrix}. \tag{6.30}$$

Since the impedance at the input of the calibration plane, $Z_{\text{IN}} = V_1/I_1$, may be determined from eqn (6.12), the terms $\overline{A_C}$, $\overline{B_C}$, and $\overline{C_C}$ may be determined by using the *OSM* procedure reported in Section 4.9, as long as three loads of known impedance, $Z_C = V_C/I_C$, are connected at the coaxial plane.

Once, $\overline{A_C}$, $\overline{B_C}$, and $\overline{C_C}$ are known, let us now analyze the connection of a PM at the coaxial plane. First, by analyzing the structure shown in Figure 6.12(b) using eqn (6.29) and (6.30), the impedance of the PM, Z_{PM}, is calculated as

$$Z_{\text{PM}} = \frac{Z_{IN,PM} - \overline{B_C}}{\overline{A_C} - Z_{IN,PM}\overline{C_C}}, \tag{6.31}$$

where $Z_{IN,PM}$ is the impedance at the input of the calibration plane, when a PM is connected at the coaxial plane (Figure 6.11).

Then, from the analysis of the structure shown in Figure 6.12(b), using eqn (6.29) and (6.30), the following expression may be derived:

$$V_1 = D_C \left(\overline{A_C} + \overline{B_C} Y_{\text{PM}}\right) V_{\text{PM}}, \tag{6.32}$$

where V_{PM} represents the voltage at the PM and $Y_{\text{PM}} = 1/Z_{\text{PM}}$. Now, let $P_{\text{PM}} = |V_{\text{PM}}|^2 \operatorname{Re}(Y_{\text{PM}})$ be the power measured in the PM. From eqn (6.32), one can derive the following expression:

$$|V_1|^2 = |D_C|^2 \left|\overline{A_C} + \overline{B_C} Y_{\text{PM}}\right|^2 \frac{P_{\text{PM}}}{\operatorname{Re}(Y_{\text{PM}})}, \tag{6.33}$$

where the term $|D_C|^2$ is calculated by assuming that the network represented by the matrix $\mathbf{T}_{\text{C}}$ is reciprocal (the determinant of $\mathbf{T}_{\text{C}}$ equals the unity) as

$$|D_C|^2 = \frac{1}{\left|\overline{A_C} - \overline{B_C C_C}\right|^2}. \tag{6.34}$$

Finally, the value of the terms $|D_X|^2$ and $|D_Y|^2$ are calculated by taking advantage of the thru connection made at the calibration reference plane, as presented next. Since a thru connection at the calibration plane is considered, the identity $V_1 = V_2$ holds. Hence, by solving eqn (6.25) and (6.28) for $|D_X|^2$ and $|D_Y|^2$, and substituting eqn (6.33) in the resultant expressions, one has

$$|D_X|^2 = \frac{|Z|^2}{|D_C|^2} \frac{\left|(1+\overline{C_X}) - (\overline{B_X} + \overline{A_X})Y_1^U\right|^2 \operatorname{Re}(Y_{\text{PM}}) \left|V_1^U\right|^2}{|\Delta_X|^2 \left|(\overline{A_C} Z_{\text{PM}} + \overline{B_C})\right|^2 P_{\text{PM}}}, \tag{6.35}$$

$$|D_Y|^2 = \frac{|D_C|^2}{|Z|^2} \frac{\left|(\overline{A_C} Z_{\text{PM}} + \overline{B_C})\right|^2 P_{\text{PM}}}{\left|(\overline{C_Y} - \overline{A_Y}) - (1 - \overline{B_Y})\right|^2 Y_2^U \operatorname{Re}(Y_{\text{PM}}) \left|V_2^U\right|^2}. \tag{6.36}$$

Once the power levels at the DUT ports have been determined, the LP measurement system is fully calibrated. Then, for each impedance synthesized by the load tuner, importance figures of merit such as output power, and large signal gain along with input and load impedance may be determined.

Other important figures of merit such as drain efficiency

$$\eta_D = P_{\text{OUT}}/P_{\text{DC}} \tag{6.37}$$

and power added efficiency

$$\text{PAE} = (P_{\text{OUT}} - P_{\text{IN}})/P_{\text{DC}} \tag{6.38}$$

may be calculated as well, as long as the direct current (DC) power dissipated in the device, P_{DC}, is measured using DC voltage and current meters. Contours of constant output power (P_{OUT}), large-signal gain (G_V, G_I, G_D, or G_P) or efficiency (η_D or PAE) or some other metric of linearity may be formed.

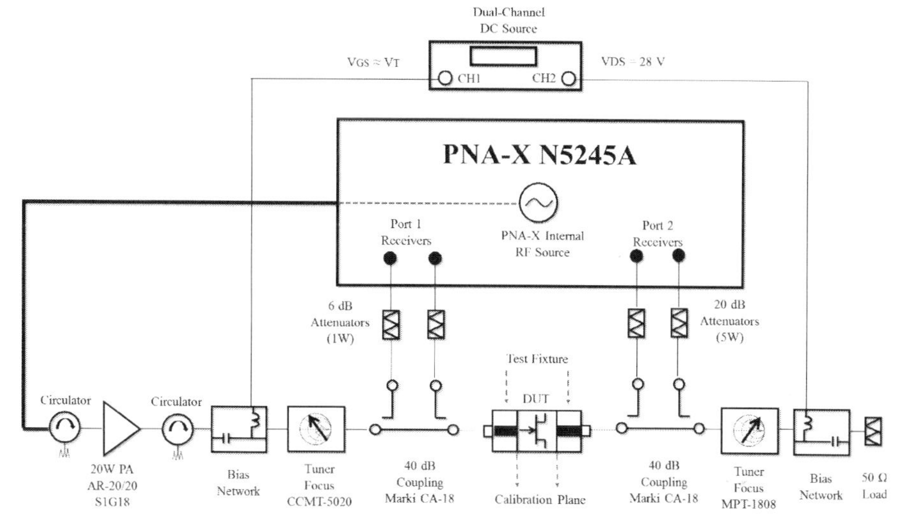

Figure 6.13 Measurement setup of the implemented passive real-time LP system.

6.5 Implementation of a Vector-Receiver LP Measurement System

The vector-receiver LP measurement system depicted in Figure 6.13 was implemented and a 10 W GaN-HEMT packaged transistor, *Cree, Inc.* CGH40010F [12], was characterized at 3.5 GHz.

As vector-receiver, the *Agilent Technologies* PNA-X N5245A was used, this instrument having external access to the instrument receivers. As RF signal source, the internal source of the PNA-X was used; alternatively, an external source may be used as long as it is phase-locked to the vector receiver. The signal emerging from the RF source is boosted by a high-power amplifier in order to drive the DUT. Circulators set at the input and output of the driver amplifier are used to protect the RF source and amplifier from possible reflections at their outputs.

Strictly, for performing an LP characterization, it is not mandatory to use an impedance tuner at the input of the DUT. Nevertheless, an impedance tuner at the DUT's input may be used to match the transistor's input to the source in order to allow maximum power transfer from the source to the transistor. It allows reducing the power from the RF source and driver amplifier required to saturate the DUT.

An important aspect to be considered when setting up an LP measurement system is the *power budget* [7]. An estimation of the power level that will

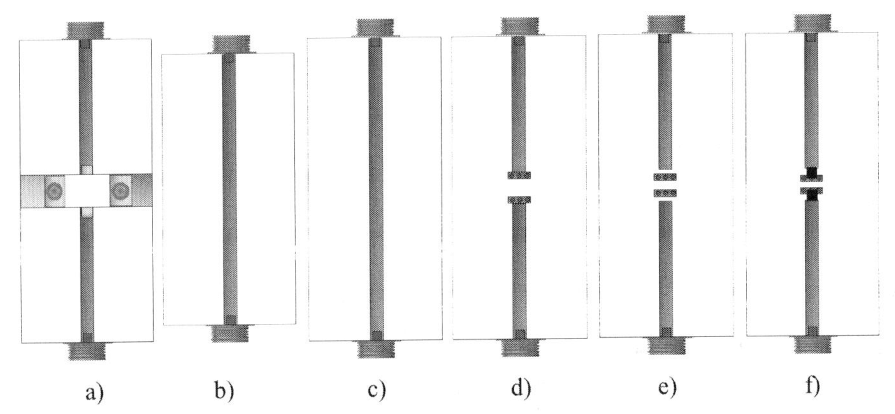

Figure 6.14 (a) Test fixture and calibration structures used to implement the TRL, TRM, TRRM, and LZZ techniques: (b) thru, (c) 50-Ω transmission line, (d) short, (e) open, and (Sf) 50-Ω load.

be managed in the whole measurement system is used to determine the characteristics of some elements of the setup. Dual directional couplers (40-dB coupling and 0.2-dB insertion loss at 3.5 GHz) were used to allow the vector-receiver to measure uncalibrated incident and reflected waves at the DUT's input and output ports. Attenuators may be placed between receivers' and couplers' arms to protect the receivers from power levels beyond their safe power limits. High-coupling directional couplers are useful to allow the receivers to operate in their linear dynamic range. Nevertheless, high-coupling could not be beneficial whether very low-power signals are required to be measured, since low-power signals entering the receivers can be affected by noise floor.

The load at the output of the DUT is varied through the use of a mechanical impedance tuner. Low-loss couplers [87] are useful to enhance the impedance tuning capabilities of the overall LP system [89]. The insertion loss of directional couplers are not a concern whether active tuning systems are used.

The DUT is mounted in a microstrip test fixture comprising 50 Ω transmission lines and SMA-to-microstrip adapters, as depicted in Figure 6.14(a). The DUT is biased by using a dual-channel DC voltage source through bias networks located at the DUT's input (gate) and output (drain) sides.

The drain voltage was set at 28 V and the gate voltage was set at a voltage level close to the transistor's threshold voltage V_T, thus operating in Class-B mode.

The measurement system was calibrated by using the calibration procedure presented in the preceding section. Since different calibration techniques (TRL, TRM, TRRM, and LZZ) were used to determine the error terms of the relative calibration part of the calibration procedure, the set of microstrip structures depicted in Figure 6.14 (b)-(f) were fabricated in a substrate RO4003 from *Rogers Corporation* [78].

6.5.1 Load-pull contours of constant power, efficiency, and gain

First, the accuracy of the calibration of the measurement system is verified. A commonly used form to verify the accuracy in the calibration of an LP system is to measure the large-signal gain of the through connection of the calibration planes (thru) [25, 10, 88], which to be consistent with the theory has to be identical to 1 (0 dB). The power gain is the commonly used parameter for this purpose; nevertheless, there are different forms to define the large-signal gain, namely the ratio of the transmitted to incident waves G_d, voltage gain G_v, current gain G_i, and power gain G_p.

Figure 6.15 shows the magnitude of the measured G_d, G_v, G_i, and G_p as a function of the magnitude of the synthesized load reflection coefficient, $\Gamma_{\text{LD}} = (Z_{\text{LD}} - Z_0)/(Z_{\text{LD}} + Z_0)$. Note that in all the cases, as the value of $|\Gamma_{\text{LD}}|$ increases, the difference between the magnitude of the large-signal gain and the expected gain also increases. This is a result commonly observed in practice [25, 86]. According to several works reported in the literature, it occurs due to uncertainty and residual errors in the relative calibration procedure (i.e., the calculation of the seven calibration terms) and several efforts have been made to mitigate them [10, 88, 4]. From the author's perspective, the origin of this result may simply be the fact that the measurement system is calibrated under low Γ_{LD} conditions (i.e., during the relative calibration procedure, the impedance loading the measurement ports is identical to Z_0) and not under high Γ_{LD} conditions.

Note that G_p may be calculated either from (6.17) using information provided by the relative calibration (vector measurements) or as the ratio $P_{\text{OUT}}/P_{\text{IN}}$ from eqn (6.25) and (6.26) using information provided by the power calibration (power measurements). To be consistent, both results have to be identical; Figure 6.15(d) shows the high correlation between these results.

Once the system is accurately calibrated, a set of impedances distributed along the Smith chart were synthesized using the LP measurement setup depicted in Figure 6.13 and the large-signal behavior of the DUT ($V_{\text{GS}} = -2.6$

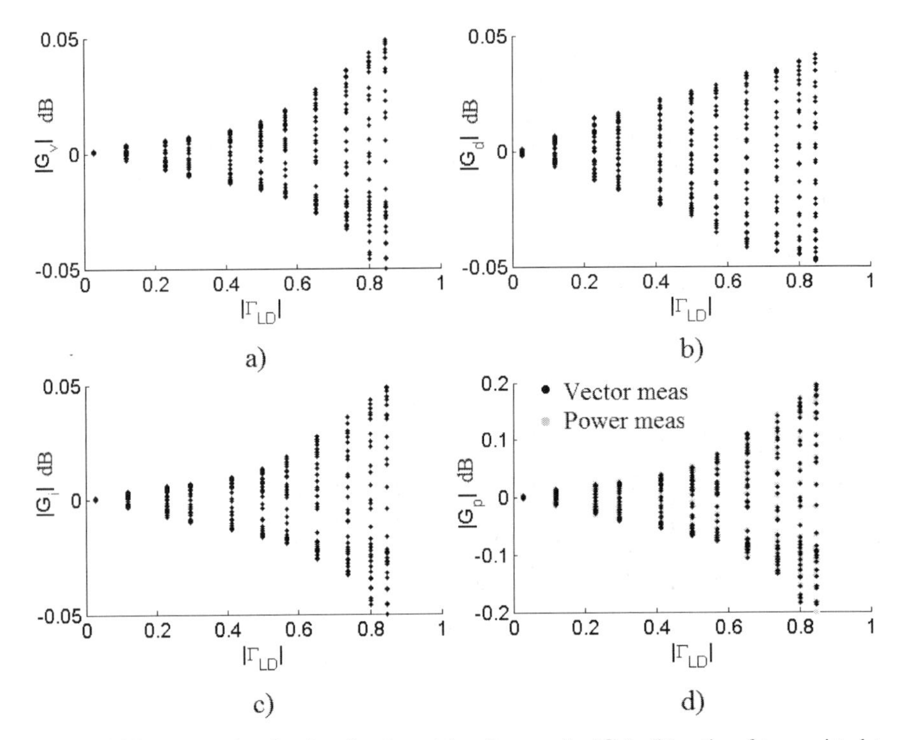

Figure 6.15 Large-signal gain of a thru: (a) voltage gain (G_v), (b) ratio of transmitted to incident waves (G_d), (c) current gain (G_i), and d) power gain (G_p), as a function of the magnitude of the load reflection coefficient Γ_{LD}.

V; $V_{\mathrm{DS}} = 28$ V) was measured. Measured LP contours of constant output power (P_{OUT}), drain efficiency (η_D), power added efficiency (PAE), and large-signal gain (G_d) are shown in Figure 6.16. The location in the Smith chart of the calculated input impedance Z_{IN} is also shown.

Note that the impedances for maximum P_{OUT}, η_D, PAE, and G_d belong to a set of load impedance values synthesized using the passive tuner. Therefore, the use of an active LP approach is not strictly required to characterize this DUT at this frequency. Figure 6.17 shows a set of LP contours of constant output power obtained using a passive LP system shown in Figure 6.13 and those obtained using an active closed-loop LP system. For the transistor used as DUT, the use of an active LP system is not mandatory since those impedances generated by the active system, which cannot be generated using the passive system, cover an area of the Smith chart that gives not relevant information in the measurement of the contours of constant P_{OUT}.

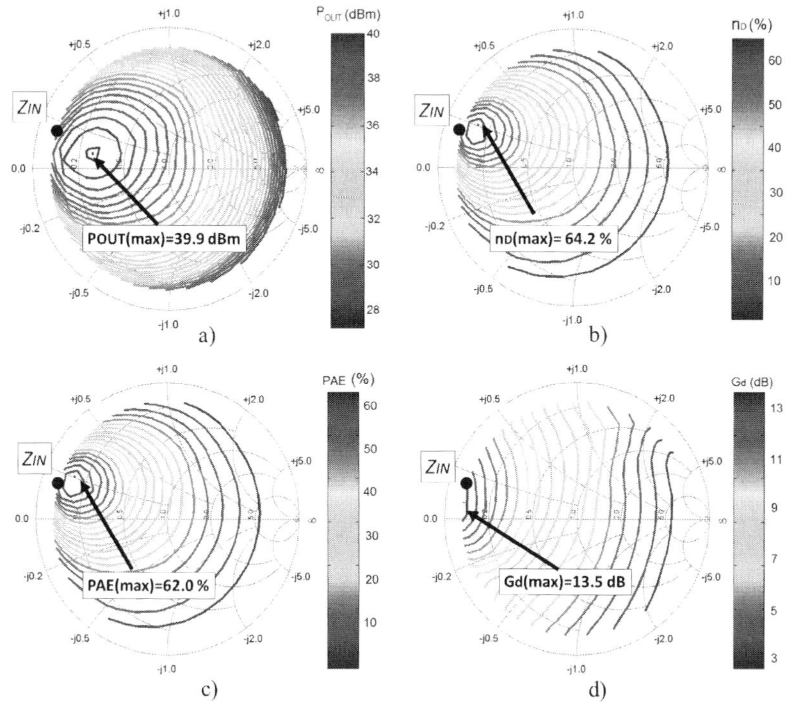

Figure 6.16 Measured LP contours of constant: (a) output power, (b) drain efficiency, (c) power added efficiency, and (d) large-signal gain of the transistor used as DUT ($V_{GS} = -2.6$ V; $V_{DS} = 28$ V).

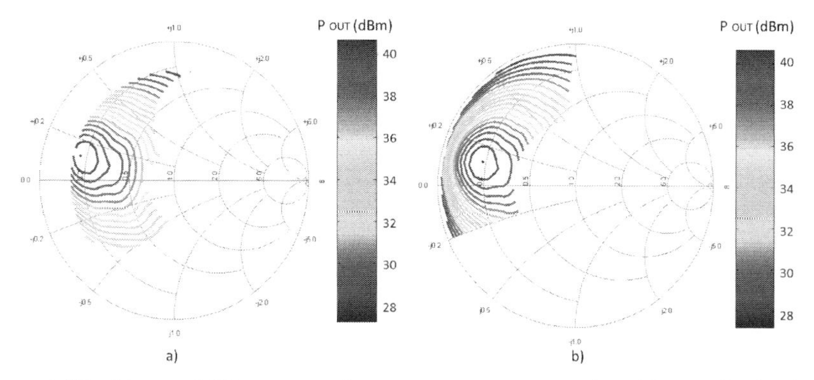

Figure 6.17 Measured LP contours of constant output power using: (a) a passive load-pull system and (b) using an active load-pull system of the transistor used as DUT ($V_{GS} = -2.6$ V; $V_{DS} = 28$ V).

6.5.2 Large-signal gain under optimal loading condition

An important parameter to consider in the large-signal characterization of a power transistor is its large-signal gain under optimal loading condition, which is a complex quantity. The information provided by the magnitude of the large-signal gain is useful to identify the power level at which the transistor is operating in its nonlinear regime through its 1-dB compression point [14]. The angle of the large-signal gain gives an insight into the phase distortion of the device as a function of the input power. Magnitude and angle of the large-signal are directly related to the *AM–AM* and *AM–PM* distortion of the transistor and, as the LP contours of constant $P_{\rm OUT}$ or η_D, are a parameter commonly used to validate non linear models of power transistors.

In Figure 6.18(a) the magnitudes of the voltage gain, current gain, power gain, and gain expressed as the ratio of transmitted to incident waves, as a function of the input power, are shown. It is noted that the $|G_p|$ is greater than $|G_i|$, $|G_v|$, and $|G_d|$. This is an expected result since these quantities are related as

$$|G_P| = |G_i|^2 \frac{\mathrm{Re}\,(Z_{\rm LD})}{\mathrm{Re}\,(Z_{\rm IN})} = |G_v|^2 \frac{\mathrm{Re}\,(Z_{\rm IN})}{\mathrm{Re}\,(Z_{\rm LD})} = |G_d|^2 \frac{(1 - |\Gamma_{\rm LD}|^2)}{(1 - |\Gamma_{\rm IN}|^2)}, \quad (6.39)$$

and, as shown in Figure 6.16, $\mathrm{Re}\,(Z_{\rm IN}) < \mathrm{Re}\,(Z_{\rm LD})$ and $|\Gamma_{\rm IN}| > |\Gamma_{\rm LD}|$. The angle of G_v, G_i, and G_d is shown in Figure 6.18(b); this information

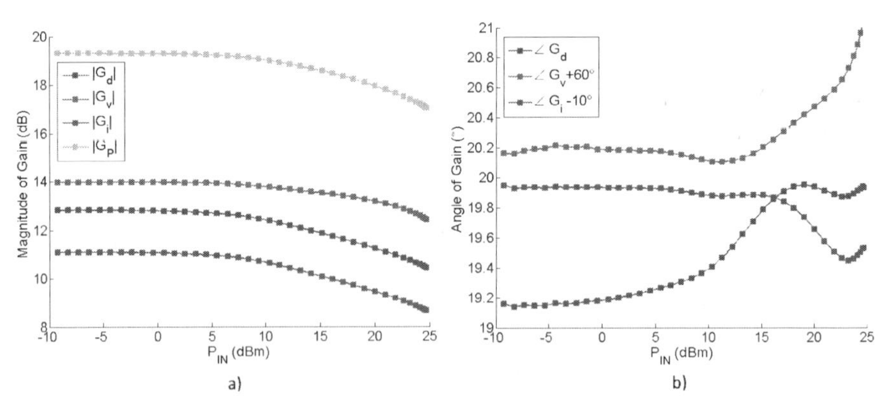

Figure 6.18 Measured large-signal gain as a function of input power of the transistor used as DUT ($V_{\rm GS} = -2.6$ V; $V_{\rm DS} = 28$ V): (a) magnitude and (b) angle.

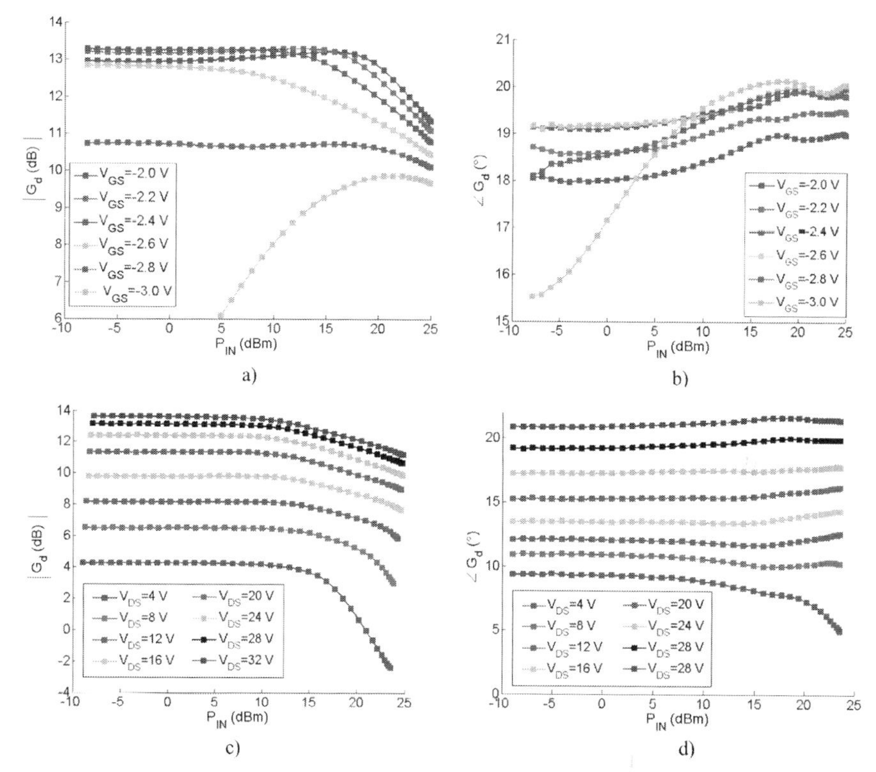

Figure 6.19 Measured large-signal gain for different bias conditions: (a) magnitude and (b) angle of $|G_d|$ as a function of V_{GS}, and (c) magnitude and (d) angle of $|G_d|$ as a function of V_{DS}.

is useful when two or more devices are required to be combined with a determined phase relationship (e.g., Doherty amplifiers, outphasing amplifiers, and balanced amplifiers).

As the optimum load impedance, the large-signal gain of a power transistor is a bias-dependent quantity. Figure 6.19 shows the magnitude and angle of G_d for different bias conditions. The dependence of G_d on the gate–source voltage (V_{GS}) is depicted in Figure 6.19(a) and (b) and its dependence on the drain–source voltage (V_{DS}) is shown in Figure 6.19(c) and (d). The information provided by these measurements may be useful whether two or more devices operating under different bias conditions are required to be combined.

Whereas the dependence of the magnitude and the angle of G_d on V_{DS} and P_{IN} obeys a well-defined behavior (i.e., the greater the V_{DS}, the greater the value of $|G_d|$ and $\angle G_d$), the dependence of the magnitude and phase of G_d on V_{GS} and P_{IN} seems to obey a more complex behavior. The large-signal gain is a technology-dependent parameter; modeling the large-signal gain of GaN-based devices by means of behavioral models [96] is a very complex task and is still an open research topic.

6.6 Evaluation of the Impact of Calibration Techniques on the LP Characterization of Microwave Transistors

In this section, the impact of different calibration techniques (TRL, TRM, TRRM, LZZ) on the LP characterization of the microwave power transistor used as DUT in the preceding section is evaluated.

6.6.1 The impact of assuming the line/load used as calibration structures as a non-reflecting element

In order to determine the most important parameters in LP, it is required to use three different types of data:

1. Uncalibrated ratios of traveling waves and absolute magnitude of traveling waves, which are measured with high degree of accuracy by using a VR.
2. Seven calibration terms partially describing the two-port networks connecting the DUT ports to the VR ports. These seven terms are obtained from the measurement of the calibration structures of a determined calibration procedure and, as shown in Chapter 4, their calculation requires minimal knowledge of the electrical characteristics of the calibration structures.
3. The impedance of one of the calibration structures used in a determined calibration technique: the characteristic impedance of a transmission line in the TRL and LZZ, the impedance of a pair of loads in the TRM, or the impedance of a one-port load in the TRRM. These parameters that depend on the type of calibration structures are generally frequency-dependent and their values have to be known prior to the calibration.

In this order, from the author's perspective, the most important source of errors in the LP characterization of a transistor are errors in the knowledge of the impedance of the transmission line/loads used in the calibration procedure. Let us define a metric $E_{Z_{LD}}$ to evaluate the impact of knowing these

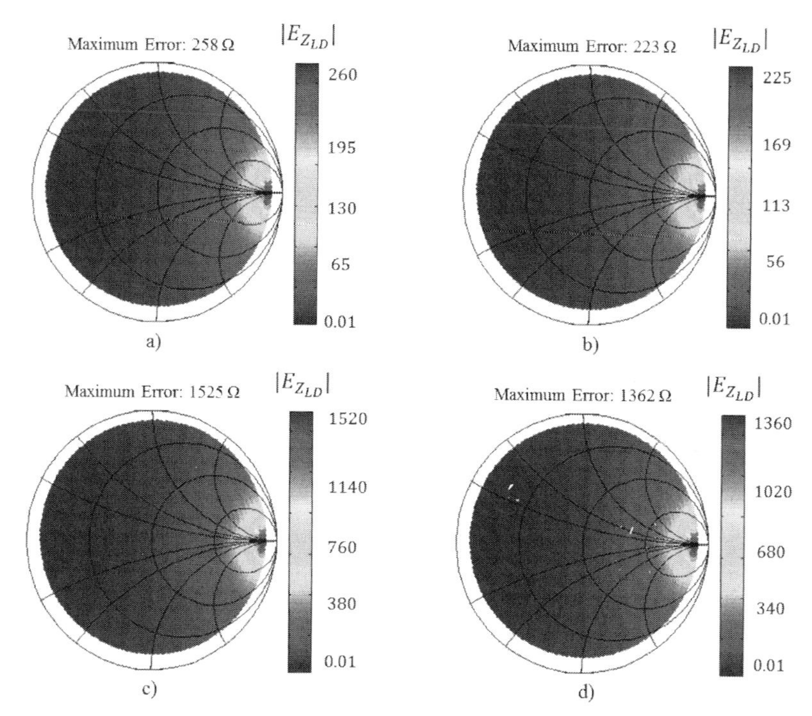

Figure 6.20 Error in the calculation of Z_{LD} due to assuming the line/load used as calibration structure in the (a) TRL, (b) LZZ, (c) TRM, and (d) TRRM calibrations as non-reflecting elements.

parameters on the determination of the LP contours. $E_{Z_{\mathrm{LD}}}$ is defined as

$$E_{Z_{\mathrm{LD}}} = Z_{\mathrm{LD}}^{\mathrm{act}} - Z_{\mathrm{LD}}^{\mathrm{err}}, \tag{6.40}$$

where $Z_{\mathrm{LD}}^{\mathrm{act}}$ is the load impedance Z_{LD} calculated using the actual value of the impedance of the line/loads used in a calibration technique and $Z_{\mathrm{LD}}^{\mathrm{err}}$ is the value of Z_{LD} calculated using an erroneous value of these parameters.

In the TRL and LZZ techniques, the same 8-mm-long transmission line of characteristic impedance $Z_L = 52.5 - j1.5\ \Omega$ was used to determine the value of $Z_{\mathrm{LD}}^{\mathrm{act}}$. In the TRM and TRRM techniques, the same load of impedance $Z_M = 53.5 + j14.0\ \Omega$ was used to determine $Z_{\mathrm{LD}}^{\mathrm{act}}$; in the TRM, the load is assumed as symmetrical.

In all the cases, $Z_{\mathrm{LD}}^{\mathrm{err}}$ was considered as the value of Z_{LD} calculated by assuming that the impedance of the line/loads used as calibration structures is purely real and identical to Z_0. Figure 6.20 shows the absolute value of the calculated $E_{Z_{\mathrm{LD}}}$ as a function of Z_{LD} for different calibration techniques.

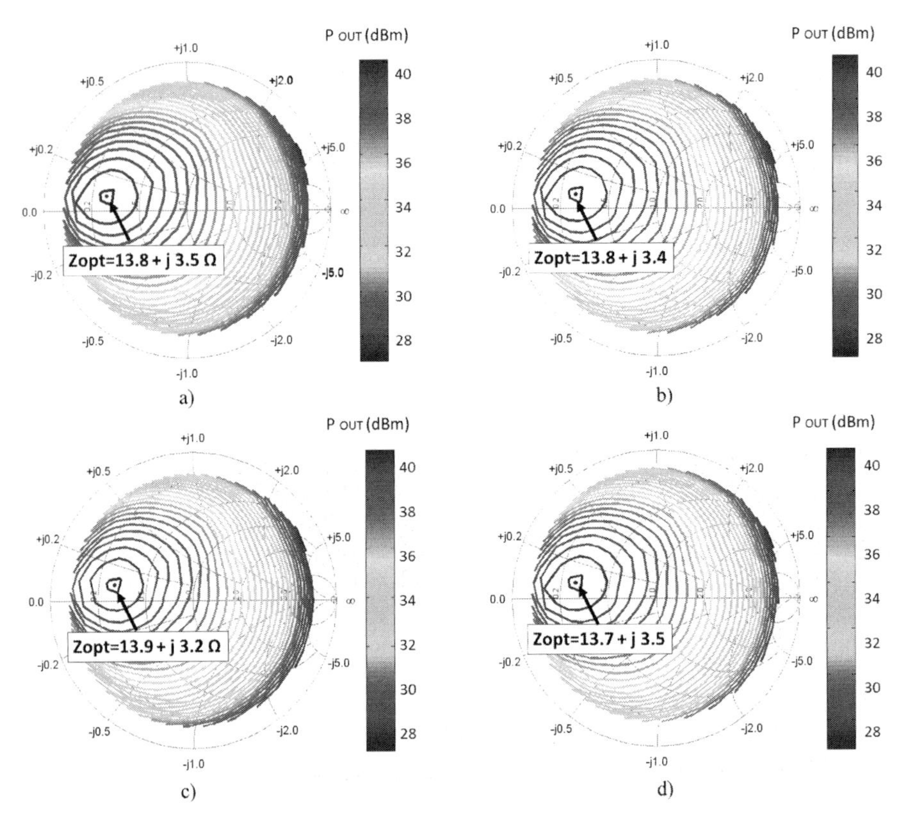

Figure 6.21 Measured LP contours of constant: (a) output power, (b) drain efficiency, (c) power added efficiency, and (d) large-signal gain of the transistor used as DUT ($V_{\mathrm{GS}} = -2.6$ V; $V_{\mathrm{DS}} = 28$ V).

An important result to note is the fact that $|E_{Z_{\mathrm{LD}}}|$ in all the cases is less in the left-hand side of the Smith chart; this is expected since the left-hand side of the Smith chart covers impedances of lower value than those in the right-hand side of the Smith chart. Power transistors manage high output current levels; therefore, power transistors require low output impedance values for optimum operation; it implies that errors in the calculation of the LP contours will be less in the vicinity of the optimum load impedance Z_{opt} than in the area of the Smith chart far from this impedance.

As can be observed in Figure 6.20, $|E_{Z_{\mathrm{LD}}}|$ is greater in the TRM and TRRM. It occurs due to the fact that the difference between Z_M and Z_0 is greater than the difference between Z_L and Z_0. It suggests that, for these calibration structures, assuming the line as a non-reflecting element

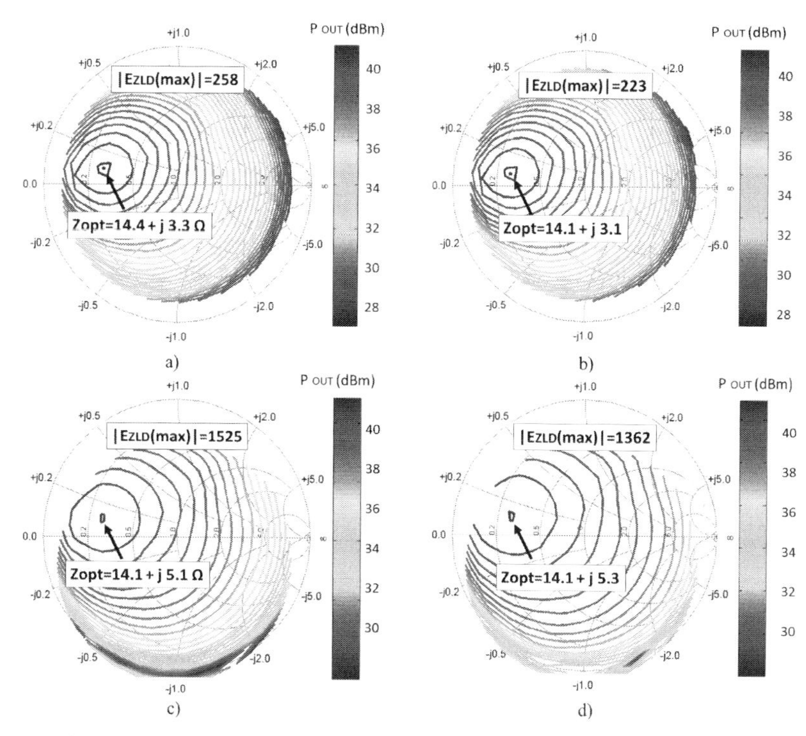

Figure 6.22 Measured LP contours of constant: (a) output power, (b) drain efficiency, (c) power added efficiency, and (d) large-signal gain of the transistor used as DUT ($V_{\mathrm{GS}} = -2.6$ V; $V_{\mathrm{DS}} = 28$ V).

introduces less errors in the calculation of Z_{LD} than assuming Z_M as a non-reflecting element. The case in which the impedance of the lines used as calibration structures greatly differs from Z_0 is further analyzed in Section 6.6.3.

Figure 6.21 shows the measured LP contours of constant output power calculated using the TRL, TRM, TRRM, and LZZ techniques in the calibration procedure. In the TRL and LZZ techniques, the actual value of Z_L was used; meanwhile, in the TRM and TRRM techniques, the actual value of Z_M was used. High correlation between the results obtained using all the aforementioned calibration techniques may be observed.

Figure 6.22 shows the measured LP contours of constant P_{OUT} calculated by assuming Z_L and Z_M as non-reflecting elements in the calibration procedure. As predicted, errors due to assuming the value of $Z_M = Z_0$ are greater than errors due to assuming the value of $Z_L = Z_0$; moreover, it can

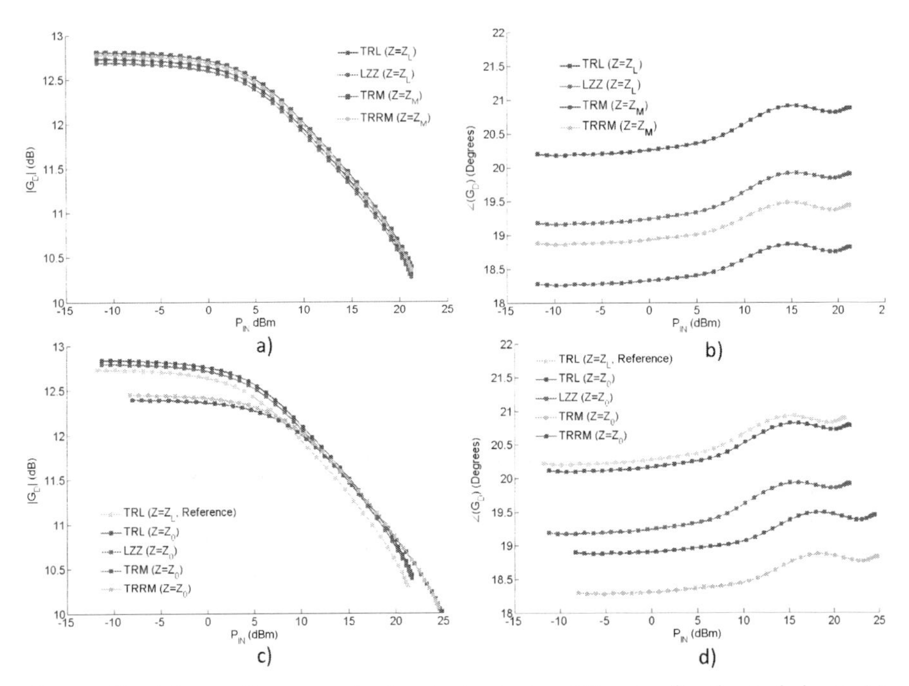

Figure 6.23 Measured large-signal gain calculated using different calibration techniques: (a) magnitude and (b) phase when the correct values of Z_L and Z_M are used in the calibration. The results obtained when Z_L and Z_M are assumed as non-reflecting elements are shown in (c) magnitude and (d) phase.

be observed that the calculation of Z_{opt} is not significantly affected in all the cases, due to the fact that Z_{opt} is located in the left-hand side of the Smith chart.

In Figure 6.23, the large-signal gain G_D calculated using different calibration techniques is presented. It may be observed that, provided that the correct values of the impedance of the line (for TRL and LZZ) and load (for the TRM and TRRM) are used in the calibration, the results obtained for all the calibration techniques are very similar. The results obtained when the lines and loads used in the calibration are assumed as non-reflecting elements are also shown; the most significative impact of this assumption is observed in the calculation of $|G_D|$ using the TRM and TRRM techniques.

6.6.2 The impact of assuming the load used in the TRM calibration as a symmetrical element

In this section, the generalized theory of the TRM calibration presented in Section 4.5 is used in the calibration of an LP measurement system [66]. The DUT and test fixture used in the preceding section were used to measure LP contours of constant output power.

The LP measurement system depicted in Figure 6.13 was calibrated using the TRM technique implemented as calibration structures: a zero-length thru, a symmetrical highly reflecting load, and an asymmetrical load as *match* standard, as shown in Figure 6.24. The asymmetrical load consists of a load of 50-Ω DC resistance at port 1 and a load of 25-Ω DC resistance (two 50-Ω loads connected in parallel) at port 2; these loads were pre-characterized and their impedance were found to be $Z_{M1} = 53.2 + j13.5$ and $Z_{M2} = 24.2 + j9.8$, respectively, at 3.5 GHz.

Figure 6.25(a) shows the LP contours of constant P_{OUT} calculated using the TRL technique (used as reference), whereas the LP contours calculated using the TRM technique taking into account both the asymmetry and frequency-dependence of the loads used as *match* calibration structure (i.e., using $Z_A = Z_{M1}$ and $Z_B = Z_{M2}$ in the calibration procedure) are shown in Figure 6.25(b). High correlation between these two results may be observed, thereby validating the accuracy of the proposed procedure.

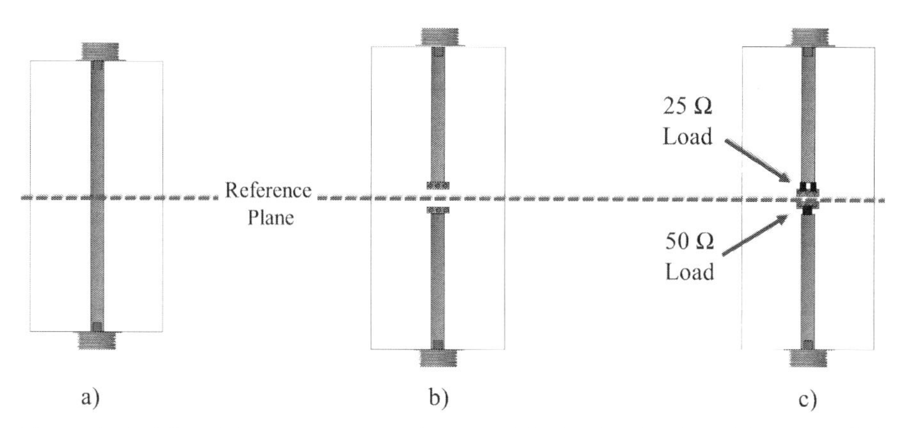

Figure 6.24 Microstrip structures used to implement the TRM calibration with asymmetrical loads: (a) thru, (b) reflecting load (short circuit), and (c) asymmetrical load. The dotted line indicates that the calibration reference plane is located at the center of the thru. Ref [66]. © 2016 IEEE. Reprinted with permission.

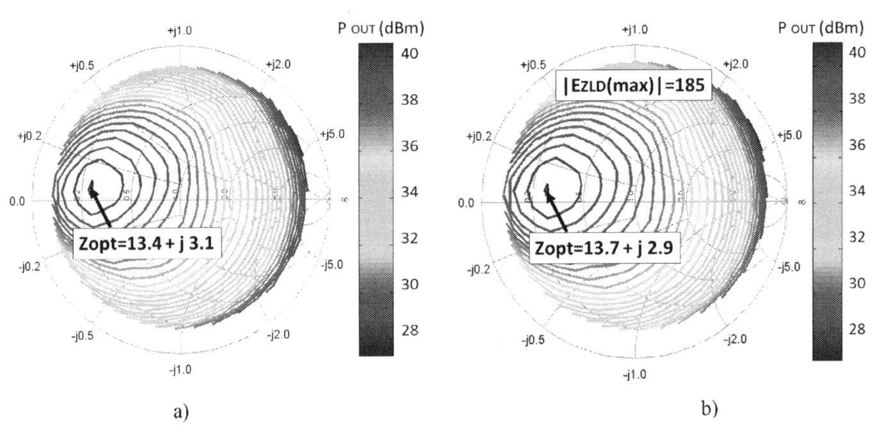

Figure 6.25 Contours of constant output power calculated using in the relative calibration: (a) TRL (used as a reference) and (b) TRM $Z_A = Z_{M1}$, $Z_B = Z_{M2}$. Ref [66]. © 2016 IEEE. Reprinted with permission.

Then, in order to show the impact of neglecting the asymmetry and/or frequency-dependence of the load used as calibration structure, the following cases were considered [66]:

1. First the frequency-dependence of the loads used as *match* standard is neglected, whereas its asymmetry is still taken into account. Figure 6.26(b) shows the LP contours of constant P_{OUT} calculated by using the DC resistance of the asymmetrical loads in the calibration procedure ($Z_A = 50$ and $Z_B = 25$). These contours greatly differ from the contours shown in Figure 6.26(a), which were calculated by using the actual value of these loads in the calibration procedure ($Z_A = Z_{M1}$ and $Z_B = Z_{M2}$).

2. Then, the impact of neglecting the asymmetry of the loads is evaluated. LP contours of constant P_{OUT} calculated by assuming that the load used as calibration structure is symmetrical and of impedance identical to the load connected at port 1 ($Z_A = Z_B = Z_{M1}$) are shown in Figure 6.26(c). Meanwhile, the contours calculated by assuming that the impedance of both loads is identical to the load connected at port 2 ($Z_A = Z_B = Z_{M2}$) are shown in Figure 6.26(e).

3. Finally, both symmetry and frequency-dependence of the loads used as match standard were neglected. Figure 6.26(d) and (f) shows the LP contours calculated by using the relative calibration of the DC resistance

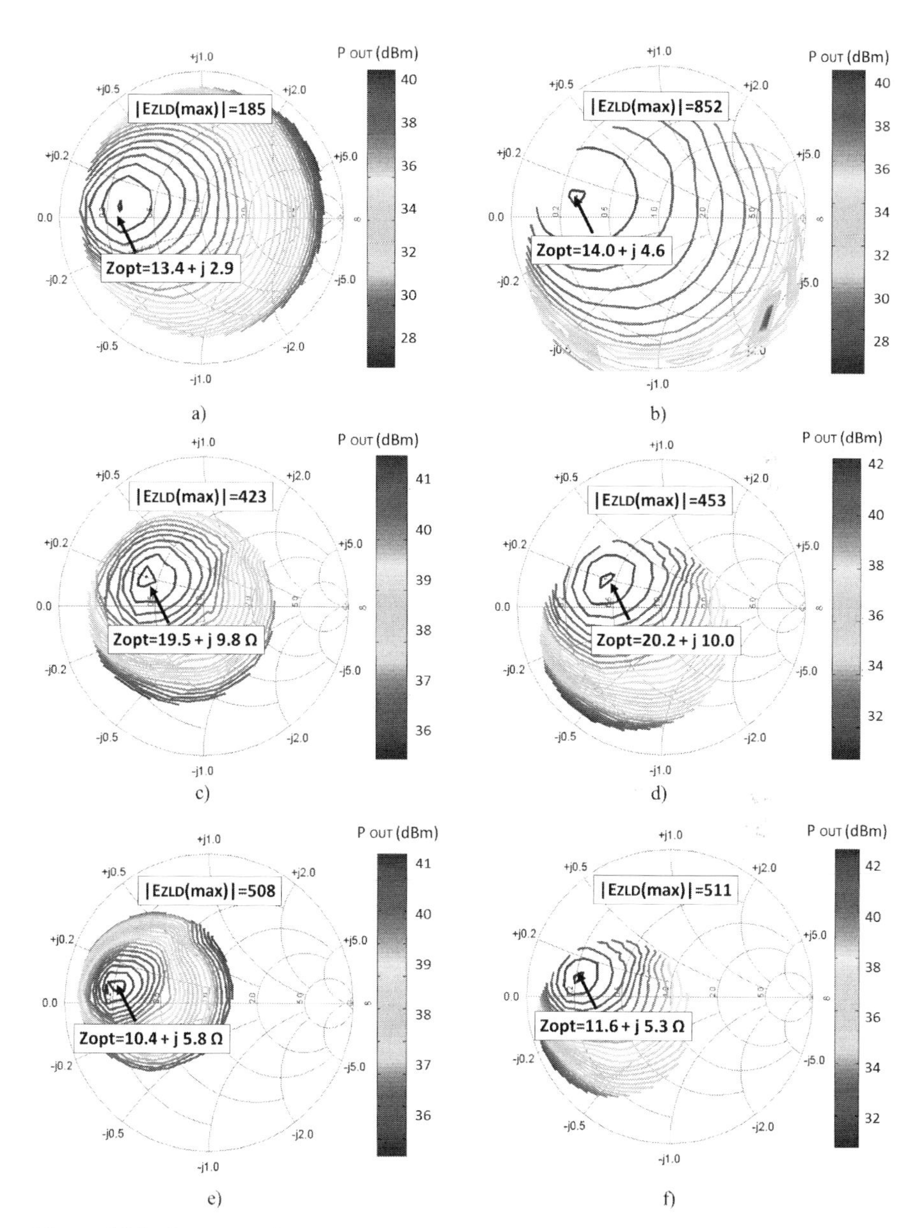

Figure 6.26 Contours of constant output power calculated using in the relative calibration: (a) TRM $Z_A = Z_{M1}$, $Z_B = Z_{M2}$, (b) TRM $Z_A = 50\ \Omega$, $Z_B = 25\ \Omega$, (c) TRM $Z_A = Z_B = Z_{M1}$, (d) TRM $Z_A = Z_B = 50\ \Omega$, (e) TRM $Z_A = Z_B = Z_{M2}$, and (f) TRM $Z_A = Z_B = 25\Omega$. Ref [66]. © 2016 IEEE. Reprinted with permission.

of the load connected at port 1 ($Z_A = Z_B = 50$) and the DC resistance of the loads connected at port 2 ($Z_A = Z_B = 25$), respectively.

From the set of results shown in Figure 6.26, it can be noted that neglecting the asymmetry of the load used as match standard causes a contraction in the size of the impedance map. This contraction is more evident when the loads are assumed to be identical to the load connected at port two than when they are assumed as identical to the load connected at port 1 in the TRM. This can be justified by inspection of eqn (6.23), which after some algebra can be expressed as

$$Z_{\text{LD}} = Z_A \cdot \frac{k - \frac{Z_B}{Z_A}}{k+1}, \tag{6.41}$$

where $k = \left(\overline{C_Y} Z_2^U - 1\right) / \left(\overline{A_Y} Z_2^U - \overline{B_Y}\right)$. In this expression, it may be noted that whether the ratio $\frac{Z_B}{Z_A} = \frac{Z_{M2}}{Z_{M1}}$ is assumed as identical to the unity, the value of Z_{LD} is underestimated, provided that $Z_{M1} > Z_{M2}$ in this example. Moreover, note that this impedance map is further reduced when the load is assumed as symmetrical and of impedance identical to the load connected at port 2. This is justified by inspection of eqn (6.41), where it may be observed that Z_{LD} is scaled by the value of Z_A.

In Figure 6.27, the voltage gain G_v calculated using the TRM calibration technique is presented. It may be observed that, provided that the correct value of the impedance of the loads used as match standards are known, the results obtained are comparable to those obtained using the reference

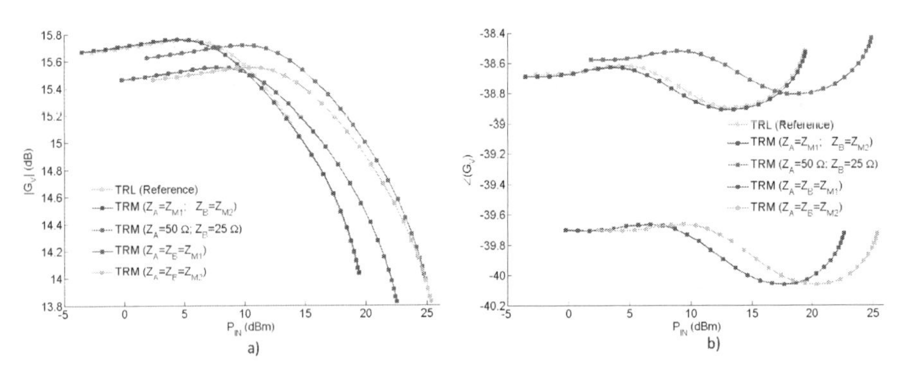

Figure 6.27 Large-signal voltage gain of the transistor used as DUT calculated using the TRM technique ($V_{\text{GS}} = -2.6$ V; $V_{\text{DS}} = 28$ V): (a) magnitude and (b) angle.

TRL technique. The results obtained when the asymmetry and/or frequency-dependence of the impedance of these structures are not taken into account are also shown.

6.6.3 The impact of knowing the impedance of the line used in the TRL on the LP characterization of power transistors

Thus far, the TRL technique implemented using lines of impedance close to Z_0 has been used in the LP characterization of packaged transistors. The width of the transmission lines used to accommodate power transistors varies according to the transistor power capabilities; the higher the power capabilities, the lower the line's impedance. Test fixtures used to mount those devices utilize wideband impedance transformers to adapt low impedance lines to non-reflecting lines [1, 72]. As a consequence, the impedance of the lines used as calibration elements in a TRL calibration also vary according to the transistor's power capabilities.

In this section, the impact of knowing the impedance of the lines used in the TRL calibration on the LP characterization of power transistors is assessed. Relevant parameters, such as input impedance, load impedance, and large-signal gain, are considered.

The LP measurement system depicted in Figure 6.13 was used to characterize a 45-W GaN-HEMT packaged transistor, *Cree, Inc.* CGH40045F [11],

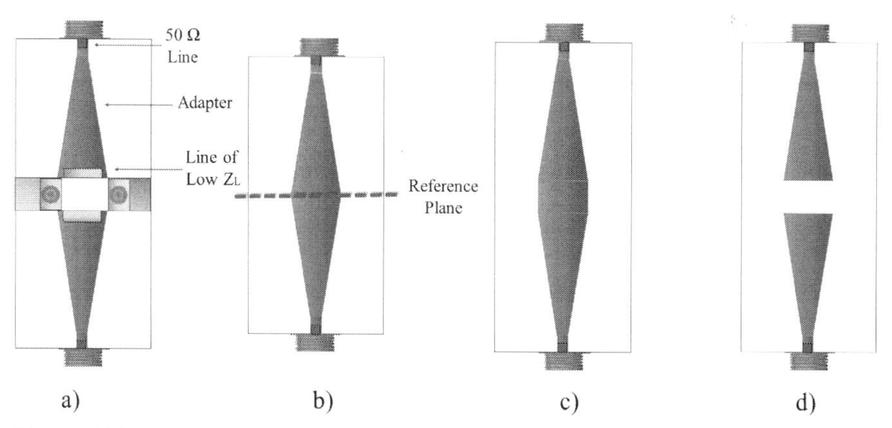

Figure 6.28 Microstrip structures used to implement the TRL calibration with lines of low impedance: (a) DUT in test fixture, (b) thru, (c) line, and (d) symmetrical reflecting load (open circuit). The dotted line indicates that the calibration reference plane is located at the center of the thru. Ref [61]. © 2015 IEEE. Reprinted with permission.

at 3.0 GHz. The DUT was mounted in a test fixture comprising wideband impedance transformers that adapt 10-Ω transmission lines to 50-Ω transmission lines, as depicted in Figure 6.28(a). The measurement setup was calibrated using the TRL procedure and the output power LP contours of the DUT (V_{DS} = 20 V; V_{GS} = −2.5 V) were measured. The TRL calibration was implemented using as calibration structures a zero-length thru, a symmetrical reflecting load (open circuit) and a transmission line of characteristic impedance close to 10 Ω, as depicted in Figure 6.28(b)-(d). In order to demonstrate in a simple manner the impact of knowing the line's impedance

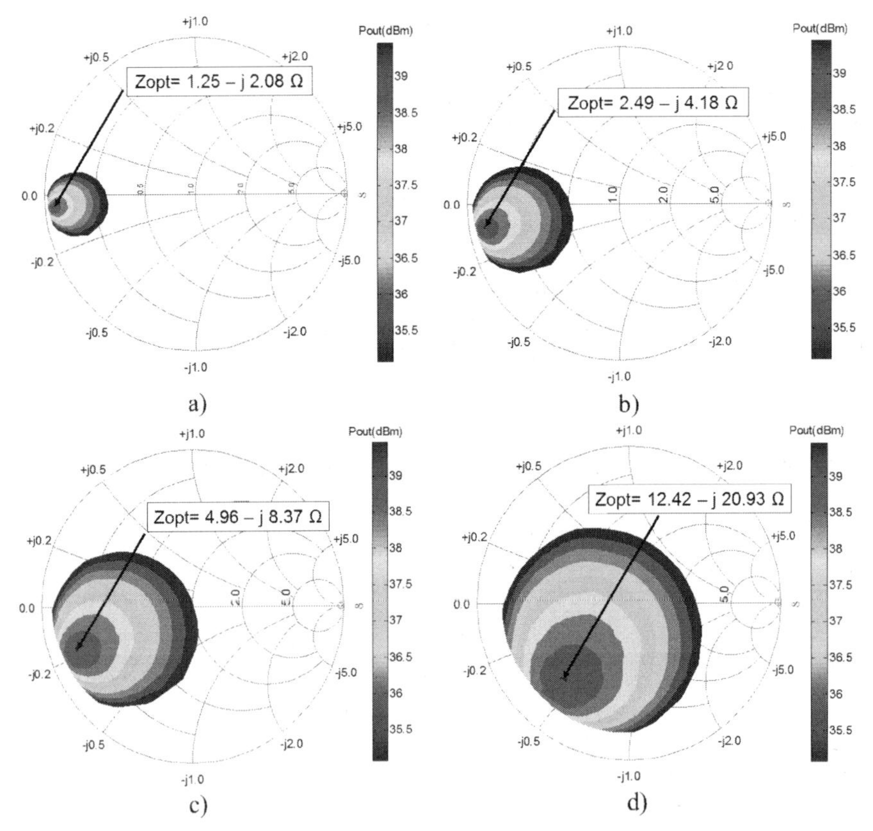

Figure 6.29 Contours of constant output power of a CGH40045 power transistor (V_{DS} = 20 V, V_{GS} = −2.5 V) using (a) $Z_L = 5$, (b) $Z_L = 10$, (c) $Z_L = 20$, and (d) $Z_L = 50$ Ω in the calibration procedure. Ref [61]. © 2015 IEEE. Reprinted with permission.

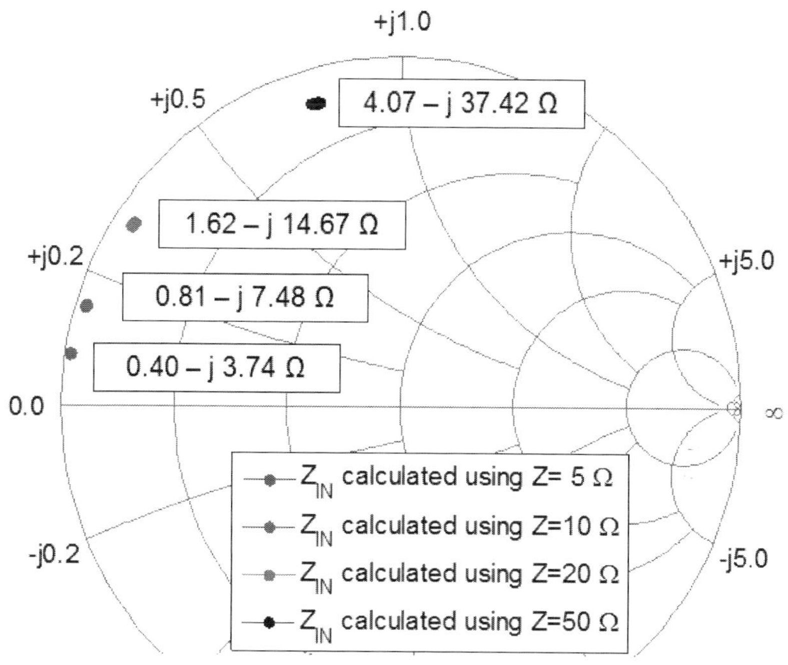

Figure 6.30 Input impedance of the DUT calculated using different values of Z in the calibration. The load impedance was fixed at $Z_{\mathrm{LD}} = Z_{opt}$. Ref [61]. © 2015 IEEE. Reprinted with permission.

on the LP characterization of the DUT, the actual value of Z_L was assumed to be real and identical to 10 Ω.

Figure 6.29 shows the output power LP contours calculated by using the actual value of Z_L (10 Ω) in the calibration procedure along with the contours calculated using three erroneous values of Z_L. It may be observed that knowing Z_L is very important in the calculation of the load impedance; an error in the knowledge of Z_L represents a proportional error in the calculation of Z_{LD}. In this order, the optimum impedance (Z_{opt}) is underestimated or overestimated, depending on the error in the value of Z_L. This parameter is very important in the design of PAs since calculating erroneously the value of (Z_{opt}) will result in suboptimum behavior of the transistor in a PA.

Then, the load impedance is fixed at Z_{opt} and the input impedance was measured. Figure 6.30 shows the input impedance calculated by using the actual value of Z_L in the calibration procedure along with the input impedance calculated using three erroneous values of Z_L. It can be observed

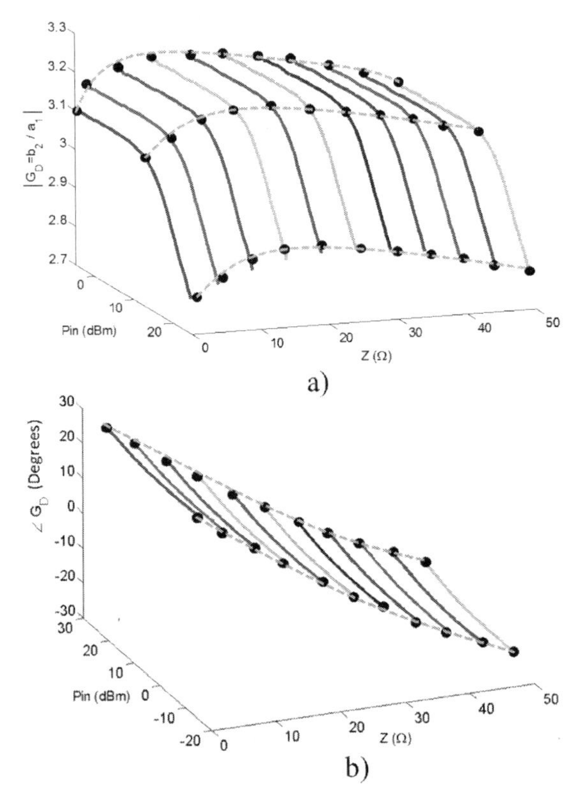

Figure 6.31 Gain of the DUT calculated using different values of Z in the calibration: (a) magnitude and (b) phase. The load impedance was fixed at $Z_{\mathrm{LD}} = Z_{opt}$. Ref [61]. © 2015 IEEE. Reprinted with permission.

that the calculated value of Z_{IN} varies proportionally with the value used for Z_L in the calibration procedure. Z_{IN} is of paramount importance in the design of PAs since it provides PA designers with the information necessary to appropriately design the input matching networks.

Finally, the impact of an error in the knowledge of Z_L in the calculation of the large-signal gain was investigated. Since the calculation of the voltage gain and the current gain is not dependent on Z_L, the gain expressed as the ratio of the transmitted to the incident waves G_d was considered.

As presented in Section 6.4, the calculation of G_D depends on Z_L through its dependence on Z_{IN} and Z_{LD}. It does not vary proportionally with Z_L as in the case of Z_{IN} and Z_{LD}. Figure 6.31 shows the magnitude and phase of G_d calculated using different values of Z_L in the calibration procedure. It may be observed that the calculation of $|G_D|$ does not vary significantly when the value of Z_L is varied. Even if the value of Z_L is considered as 1 Ω or as 50 Ω, the calculation of $|G_D|$ varies less than 10% from the $|G_D|$ calculated using the actual value of Z_L. The calculation of the phase of G_D, on the other hand, is greatly affected by errors in the knowledge of the Z_L.

A

Diagonalization of the Matrix Representation of a Uniform Transmission Line

A.1 Diagonalization of the ABCD-parameters Matrix Representation of a Uniform Transmission Line

A uniform transmission line of characteristic impedance Z_L, propagation constant γ, and length l can be represented in ABCD-parameters formalism as

$$\mathbf{L_A} = \begin{bmatrix} \cosh(\gamma l) & Z_L \sinh(\gamma l) \\ Z_L^{-1} \sinh(\gamma l) & \cosh(\gamma l) \end{bmatrix}. \tag{A.1}$$

Distinct eigenvalues are a necessary condition for that $\mathbf{L_A}$ to be diagonalizable. Then the first step is calculating the eigenvalues of the matrix $\mathbf{L_A}$. Since the determinant of $\mathbf{L_A}$ equals the unity, its eigenvalues (λ_1 , $\frac{1}{\lambda_1}$) are reciprocal to each another (λ_2=$\frac{1}{\lambda_1}$). The eigenvalues of $\mathbf{L_A}$ are determined by calculating the determinant of the matrix

$$(\mathbf{L_A} - \lambda \mathbf{I})\mathbf{x} = \mathbf{0}, \tag{A.2}$$

where λ is the vector of eigenvalues associated with $\mathbf{L_A}$, and $\mathbf{x}$ is a non-null vector referred to as the matrix of eigenvectors associated with λ.

The eigenvalue vector can be determined by calculating the determinant of $(\mathbf{L_A} - \lambda \mathbf{I})\mathbf{x} = \mathbf{0}$, which denotes the conditions in which $(\mathbf{L_A} - \lambda \mathbf{I}) = \mathbf{0}$ is singular.

$$\det(\mathbf{L_A}) = \begin{vmatrix} \cosh(\gamma l) - \lambda & Z_L \sinh(\gamma l) \\ Z_L^{-1} \sinh(\gamma l) & \cosh(\gamma l) - \lambda \end{vmatrix} = 0. \tag{A.3}$$

By developing eqn (A.3), the following equation may be derived:

$$\lambda^2 - 2\lambda cosh(\gamma L) + \left[cosh^2(\gamma L) - sinh^2(\gamma L)\right] = 0. \tag{A.4}$$

In eqn (A.4), by using $cosh^2(\gamma L) - sinh^2(\gamma L) = 1$ and solving the resultant equation for λ, the following equation may be obtained:

$$\lambda_{1,2} = cosh(\gamma L) \pm \sqrt{cosh^2(\gamma L) - 1}. \tag{A.5}$$

Then, by using the definition $cosh^2(\gamma L) - 1 = sinh^2(\gamma L)$, one obtains the following equations for the two eigenvalues:

$$\lambda_1 = cosh(\gamma L) + sinh(\gamma L) = e^{\gamma L} \tag{A.6}$$

$$\lambda_2 = cosh(\gamma L) - sinh(\gamma L) = e^{-\gamma L}. \tag{A.7}$$

Since matrix $\mathbf{L_A}$ has two distinct eigenvalues, it is diagonalizable and can be expressed as

$$\mathbf{L_A} = \mathbf{T_Z}\mathbf{T}_\lambda\mathbf{T_Z^{-1}}, \tag{A.8}$$

where $\mathbf{T}_\lambda$ is a diagonal matrix that can be expressed as

$$\mathbf{T}_\lambda = \begin{bmatrix} e^{\gamma L} & 0 \\ 0 & e^{-\gamma L} \end{bmatrix} \tag{A.9}$$

and $\mathbf{T_Z}$ is a matrix formed by two linearly independent column vectors, referred to as eigenvectors associated with $\mathbf{L_A}$. To determine $\mathbf{T_Z}$, the following two equations

$$\begin{bmatrix} \cosh(\gamma l) - \lambda_1 & Z_L\sinh(\gamma l) \\ Z_L^{-1}\sinh(\gamma l) & \cosh(\gamma l) - \lambda_1 \end{bmatrix} \begin{bmatrix} x_{11} \\ x_{21} \end{bmatrix} = \begin{bmatrix} 0 \\ 0 \end{bmatrix}, \tag{A.10}$$

$$\begin{bmatrix} \cosh(\gamma l) - \lambda_2 & Z_L\sinh(\gamma l) \\ Z_L^{-1}\sinh(\gamma l) & \cosh(\gamma l) - \lambda_2 \end{bmatrix} \begin{bmatrix} x_{12} \\ x_{22} \end{bmatrix} = \begin{bmatrix} 0 \\ 0 \end{bmatrix} \tag{A.11}$$

must be used as follows. The following expressions may be obtained by developing eqn (A.10) and (A.11):

$$[cosh(\gamma l) - \lambda_1]\frac{x_{11}}{x_{21}} = -Z_L sinh(\gamma l), \tag{A.12}$$

$$[cosh(\gamma l) - \lambda_2]\frac{x_{12}}{x_{22}} = -Z_L sinh(\gamma l). \tag{A.13}$$

Then, note that by using the following identities

$$cosh(\gamma l) - \lambda_1 = \frac{e^{\gamma l} + e^{-\gamma l}}{2} - e^{\gamma l} = -sinh(\gamma l) \tag{A.14}$$

$$cosh(\gamma l) - \lambda_2 = \frac{e^{\gamma l} + e^{-\gamma l}}{2} - e^{-\gamma l} = sinh(\gamma l) \tag{A.15}$$

in eqn (A.12) and (A.13), the following two expressions result:

$$\frac{x_{11}}{x_{21}} = Z_L, \tag{A.16}$$

$$\frac{x_{12}}{x_{22}} = -Z_L. \tag{A.17}$$

Thus, two different representations for the matrix of eigenvectors $\mathbf{T_Z}$ may be encountered as

$$\mathbf{T_Z} = \begin{bmatrix} 1 & Z_L \\ Z_L^{-1} & 1 \end{bmatrix}, \tag{A.18}$$

$$\mathbf{T_Z} = \begin{bmatrix} -Z_L & Z_L \\ 1 & 1 \end{bmatrix}. \tag{A.19}$$

A.2 Diagonalization of the T-parameters Matrix of a Uniform Transmission Line

A uniform transmission line of reflection coefficient $\Gamma_L = (Z_L - Z_0)/(Z_L + Z_0)$, propagation constant γ, and length L can be represented in T-parameters as

$$\mathbf{T_A} = \frac{1}{(1-\Gamma^2)e^{-\gamma L}} \begin{bmatrix} e^{-2\gamma L} - \Gamma_L^2 & \Gamma_L(1 - e^{-2\gamma L}) \\ -\Gamma_L(1 - e^{-2\gamma L}) & 1 - \Gamma_L^2 e^{-2\gamma L} \end{bmatrix}. \tag{A.20}$$

The matrix $\mathbf{T_A}$ may be diagonalized as long as it has two distinct eigenvalues. Then, the first step is calculating its eigenvalues. Since the determinant of $\mathbf{T_A}$ equals the unity, the eigenvalues of the matrix $\mathbf{T_A}$ are reciprocal($\Lambda_1 = 1/\Lambda_2$). The eigenvalues of $\mathbf{T_A}$ are determined by calculating the determinant of the matrix

$$(\mathbf{T_A} - \lambda \mathbf{I})\mathbf{x} = 0, \tag{A.21}$$

where λ is called an eigenvalue of the matrix $\mathbf{T_A}$, $\mathbf{x}$ is a non-null vector referred to as the eigenvector of $\mathbf{T_A}$ associated with the eigenvalue λ. The eigenvalues can be determined by calculating the determinant of $(\mathbf{T_A} - \lambda\mathbf{I})\mathbf{x} = 0$ that occurs when $\mathbf{T_A} - \lambda\mathbf{I}$ is singular. By calculating the $\det((\mathbf{T_A} - \lambda \mathbf{I})\mathbf{x} = 0)$, the following expression may be obtained:

$$\left[\frac{e^{-2\gamma L} - \Gamma_L^2}{(1-\Gamma_L^2)e^{-\gamma L}} - \lambda\right]\left[\frac{1 - \Gamma_L^2 e^{-2\gamma L}}{(1-\Gamma_L^2)e^{-\gamma L}} - \lambda\right] + \left[\Gamma_L \frac{1 - e^{-2\gamma L}}{(1-\Gamma_L^2)e^{-\gamma L}}\right]\left[\Gamma_L \frac{1 - e^{-2\gamma L}}{(1-\Gamma_L^2)e^{-\gamma L}}\right] = 0. \qquad (A.22)$$

Then, after developing eqn (A.22), the following quadratic equation may be derived:

$$\lambda^2 - \lambda(1 - e^{-2\gamma L})e^{\gamma L} + 1 = 0. \qquad (A.23)$$

Next solving the quadratic equation, the two eigenvalues may be obtained $\lambda_1 = e^{\gamma L}$, $\lambda_2 = e^{-\gamma L}$. Since the two eigenvalues are different, the matrix $\mathbf{T_A}$ may be diagonalized, and therefore expressed as

$$\mathbf{T_A} = \mathbf{T_\Gamma} \mathbf{T_\lambda} \mathbf{T_\Gamma^{-1}}, \qquad (A.24)$$

where $\mathbf{T_\lambda}$ is a diagonal matrix that is defined as

$$\mathbf{T_\lambda} = \begin{bmatrix} \lambda_1 & 0 \\ 0 & \lambda_2 \end{bmatrix}. \qquad (A.25)$$

To determine the vectors forming the matrix $\mathbf{T_\Gamma}$, the following equation must be solved for the eigenvalue λ_1:

$$\begin{bmatrix} \frac{e^{-2\gamma L} - \Gamma_L^2}{K} - \lambda_1 & \frac{\Gamma_L(1-e^{-2\gamma L})}{K} \\ \frac{-\Gamma_L(1-e^{-2\gamma L})}{K} & \frac{1-\Gamma_L^2 e^{-2\gamma L}}{K} - \lambda_1 \end{bmatrix}\begin{bmatrix} x_{11} \\ x_{21} \end{bmatrix} = \begin{bmatrix} 0 \\ 0 \end{bmatrix}, \qquad (A.26)$$

where $K = \frac{1}{(1-\Gamma^2)e^{-\gamma L}}$. Meanwhile, the following equation must be solved for the eigenvalue λ_2:

$$\begin{bmatrix} \frac{e^{-2\gamma L} - \Gamma_L^2}{K} - \lambda_2 & \frac{\Gamma_L(1-e^{-2\gamma L})}{K} \\ \frac{-\Gamma_L(1-e^{-2\gamma L})}{K} & \frac{1-\Gamma_L^2 e^{-2\gamma L}}{K} - \lambda_2 \end{bmatrix}\begin{bmatrix} x_{12} \\ x_{22} \end{bmatrix} = \begin{bmatrix} 0 \\ 0 \end{bmatrix}. \qquad (A.27)$$

By developing eqn (A.26) and (A.27), the following equations may be derived:

$$\left[(e^{-2\gamma L} - \Gamma_L^2) - \lambda_1(1 - \Gamma_L^2)e^{-\gamma L}\right] x_{11} + \left[\Gamma_L(1 - e^{-2\gamma L})\right] x_{21} = 0, \qquad (A.28)$$

$$\left[(e^{-2\gamma L} - \Gamma_L^2) - \lambda_2(1 - \Gamma_L^2)e^{-\gamma L}\right] x_{12} + \left[\Gamma_L(1 - e^{-2\gamma L})\right] x_{22} = 0. \qquad (A.29)$$

By substituting λ_1 and λ_2 in eqn (A.28) and (A.29), the following expressions may be obtained for x_{11}/x_{21} and x_{12}/x_{22}:

$$\frac{x_{11}}{x_{21}} = \Gamma_L, \tag{A.30}$$

$$\frac{x_{12}}{x_{22}} = \Gamma_L. \tag{A.31}$$

Thus, the following two definitions for the matrix T_Γ are identified:

$$\mathbf{T}_\Gamma = \begin{bmatrix} \Gamma_L & 1 \\ 1 & \Gamma_L \end{bmatrix}, \tag{A.32}$$

$$\mathbf{T}_\Gamma = \begin{bmatrix} 1 & \Gamma_L^{-1} \\ \Gamma_L^{-1} & 1 \end{bmatrix}. \tag{A.33}$$

Bibliography

[1] P. Aaen et al. *A wideband method for the rigorous low-impedance load-pull measurement of high-power transistors suitable for large-signal model validation.* 56th Automatic Radio Frequency Techniques Group (ARFTG) Conference Digest. Boulder, CO, USA. (2000), pp. 163–169.

[2] Zaid Aboush et al. *Active harmonic load-pull system for characterizing highly mismatched high power transistors.* IEEE Microwave Theory and Techniques International Symposium (MTT-IMS) Digest. Cherry Hill, NJ, USA. (2005), pp. 12–17.

[3] Zaid Aboush et al. *High power active harmonic load-pull system for characterization of high power 100-watt transistors.* 2005 European Microwave Conference. (2005), pp. 1–4.

[4] A. Aldoumani et al. *Enhanced vector calibration of load-pull measurement systems.* 83rd Automatic Radio Frequency Techniques Group Conference Digest. Tampa, FL, USA. (2014), pp. 1–4.

[5] Anritsu-Corporation. *Vector Star MS4640B Series.* Application Note 5A - 051. (2014), pp. 1–32.

[6] G.P. Bava, Umberto Pisani, and V. Pozzolo. *Active load technique for load-pull characterization at microwave frequencies.* Electronics Letters 18.4 (1982), pp. 178–180.

[7] L. Betts, D. T. Bespalko, and S. Boumaiza. *Application of Agilent's PNA-X Nonlinear Vector Network Analyzer and X-Parameters in Power Amplifier Design.* Agilent Technologies White Paper (2011).

[8] B. Bianco and M. Parodi. *Determination of the propagation constant of uniform microstrip lines.* Alta Frequence 45 (1976), pp. 107–110.

[9] Ashok Bindra. *Wide-Bandgap-Based Power Devices: Reshaping the power electronics landscape.* IEEE Power Electronics Magazine 2.1 (2015), pp. 108–116.

[10] Serena Bonino, Valeria Teppati, and Andrea Ferrero. *Further improvements in real-time load-pull measurement accuracy.* IEEE Microwave and Wireless Components Letters 20.2 (2010), pp. 121–123.

[11] Inc. Cree. *CGH40045 Power Transistor.* Data Sheet. (2015), pp. 1–16.

[12] Inc. Cree. *CGH40010 Power Transistor.* Data Sheet. (2016), pp. 1–15.

[13] S.C. Cripps. *A theory for the prediction of GaAs FET load-Pull power contours.* IEEE Microwave Theory and Techniques International Symposium (MTT-IMS) Digest. Boston, MA, USA. (1983), pp. 221–223.

[14] Steve C. Cripps. *RF Power Amplifiers for Wireless Communications.* First. Norwood, MA: Artech House, 2006. ISBN: 9780521193238.

[15] J. M. Cusack, Stewart M. Perlow, and B. S. Perlman. *Automatic load contour mapping for microwave power transistors.* IEEE Microwave Theory and Techniques International Symposium (MTT-IMS) Digest. Atlanta, GA, USA. (1974), pp. 269–271.

[16] Andrew Davidson, Keith Jones, and Eric Strid. *LRM and LRRM calibrations with automatic determination of load inductance.* 36th Automatic Radio Frequency Techniques Group (ARFTG) Conference Digest. Monterrey, CA, USA. (1990), pp. 57–63.

[17] Frédérique Deshours et al. *Experimental comparison of load-pull measurement systems for nonlinear power transistor characterization.* IEEE Transactions on Instrumentation and Measurement 2 (1997), pp. 1251–1255.

[18] Janusz A. Dobrowolski. *Introduction to Computer Methods for Microwave Circuit Analysis and Design.* Artech House., 1991. ISBN: 0890065055.

[19] Ralf Doerner and Andrej Rumiantsev. *Verification of the wafer-level LRM+ calibration technique for GaAs applications up to 110 GHz.* 65th Automatic Radio Frequency Techniques Group Conference Digest. Long Beach, CA, USA. (2005), pp. 5–9.

[20] Achim Enders. *An accurate measurement technique for line properties, junction effects, and dielectric and magnetic material parameters.* IEEE Transactions on Microwave Theory and Techniques 37 (1989), pp. 598–605.

[21] G.F. Engen and C.A. Hoer. *Thru-Reflect-Line: An Improved Technique for Calibrating the Dual Six-Port Automatic Network Analyzer.* IEEE Transactions Microwave Theory and Techniques 27.12 (1979), pp. 987–993.

[22] Hermann Eul and B. Schiek. *Thru-Match-Reflect: one result of a rigorous theory for de-embedding and network analyzer calibration.* Proceedings of 18th European Microwave Conference (EuMC) (1988), pp. 909–914.

[23] Hermann Eul and B. Schiek. *A generalized theory and new calibration procedures for network analyzer self-calibration.* IEEE Transactions on Microwave Theory and Techniques 39.4 (1991), pp. 724–731.

[24] R. F.Scholz et al. *Advanced technique for broadband on-wafer RF device characterization.* 63th Automatic Radio Frequency Techniques Group (ARFTG) Conference Digest. Fort Worth, TX, USA. (), pp. 91–96.

[25] A. Ferrero, V. Teppati, and A. Carullo. *Accuracy evaluation of on-wafer load-pull measurements.* IEEE Transactions on Microwave Theory and Techniques. 49.1 (2001), pp. 39–43.

[26] Andrea Ferrero and Marco Pirola. *Harmonic load-pull techniques: An overview of modern systems.* IEEE Microwave Magazine 14.4 (2013), pp. 116–123.

[27] Andrea Ferrero and Umberto Pisani. *Two-port network analyzer calibration using an unknown 'thru'.* IEEE Microwave and Guided Wave Letters 2.12 (1992), pp. 505–507.

[28] Andrea Ferrero and Umberto Pisani. *An improved calibration technique for on-wafer large-signal transistor characterization.* IEEE Transactions on Instrumentation and Measurement 42.2 (1993), pp. 360–364.

[29] Jim. Fitzpatrick. *Error Models for Systems Measurement.* Microwave Journal (May 1, 1978).

[30] D.A. Frickey. *Conversions between S, Z, Y, H, ABCD, and T parameters which are valid for complex source and load impedances.* IEEE Transactions on Microwave Theory and Techniques. 42.2 (1994), pp. 205–211. DOI: 10.1109/22.275248.

[31] Manuel Alejandro Pulido Gaytan. "The impact of VNA calibration techniques on the load-pull characterization of microwave power transistors". PhD thesis. Centro de Investigación Científica y de Educación Superior de Ensenada, Baja California, Mexico., 2016.

[32] Fadhel M. Ghannouchi and Mohammad Hashmi. *Load-pull techniques and their applications in power amplifiers design.* IEEE Bipolar/BiCMOS Circuits and Technology Meeting (BCTM) Digest. Atlanta, GA, USA. 2 (2011), pp. 133–137.

[33] Fadhel M. Ghannouchi and Mohammad S. Hashmi. *Load-Pull Techniques with Applications to Power Amplifier Design.* Springer, 2013. ISBN: 0890065055.

[34] Fadhel M. Ghannouchi et al. *Loop Enhanced Passive Source- and Load-Pull Technique for High Reflection Factor Synthesis.* IEEE Transactions on Microwave Theory and Techniques 58.11 (2010), pp. 2952–2959.

[35] Mike Golio. *The RF and Microwave Handbook.* CRC, 2010. ISBN: 0890065055.

[36] G. Gonzalez. *Microwave Transistor Amplifiers: Analysis and Design.* Upper Saddle River, NY: Prentice Hall, 1997. ISBN: 9780521193238.

[37] Fabien De Groote et al. *Introduction to measurements for power transistor characterization.* IEEE Microwave Magazine 9.3 (2008), pp. 70–85.

[38] H.J.Simon. *Method of calibrating a network analyzer.* U.S. Patent No. 7,782,065 (2010), pp. 1–15.

[39] R.A. Hackborn. *An Automatic Network Analyzer System.* Microwave Journal (1968).

[40] Mohammad Hashmi and Fadhel M. Ghannouchi. *Introduction to load-pull systems and their applications.* IEEE Instrumentation and Measurement Magazine 16.1 (2013), pp. 30–36.

[41] Mohammad Hashmi et al. *Highly reflective load-pull.* IEEE Microwave Magazine 12.4 (2011), pp. 96–107.

[42] Leonard Hayden. *An enhanced Line-Reflect-Reflect-Match calibration.* 67th Automatic Radio Frequency Techniques Group (ARFTG) Conference Digest. San Francisco, CA, USA. (2006), pp. 143–149.

[43] Holger Heuermann, A. Rumiantsev, and Stephen Schott. *Advanced on-wafer multiport calibration methods for mono- and mixed-mode device characterization.* 63rd Automatic Radio Frequency Techniques Group (ARFTG) Conference Digest. Fort Worth, TX, USA. (1995), pp. 91–96.

[44] M. Hiebel. *Fundamentals of Vector Network Analysis.* Rohde and Swartz., 2007. ISBN: 3939837067.

[45] J.A.Reynoso-Hernandez and Everardo Inzunza-Gonzalez. *A straightforward de-embedding method for devices embedded in test fixtures.* 57th Automatic Radio Frequency Techniques Group Conference Digest. Phoenix, AZ, USA. (2001), pp. 1–5.

[46] Keysight-Technologies. *PNA Family of Microwave Network Analyzers.* Configuration Guide. (2015), pp. 1–36.

[47] K. Kurokawa. *Power Waves and the Scattering Matrix.* IEEE Transactions on Microwave Theory and Techniques 13.2 (1965), pp. 194–202.

[48] Moon-Que Lee and Sangwook Nam. *An accurate broadband measurement of substrate dielectric constant.* IEEE Microwave and Guided Wave Letters 6 (1996), pp. 168–170.

[49] Alain M. Mangan, Sorin P. Voinigescu, and Ming-Ta Yang andMihai Tazlauanu. *De-embedding transmission line measurements for accurate modeling of IC designs.* IEEE Transactions on electron devices 53.2 (2006), pp. 235–241.

[50] Mauro Marchetti. "Mixed-SignalInstrumentationfor Large-Signal Device Characterization and Modelling". PhD thesis. Delft University of Technology, 2013.

[51] Roger B. Marks. *A multiline method of network analyzer calibration.* IEEE Transactions on Microwave Theory and Techniques 39.7 (1992), pp. 92–96.

[52] Roger B. Marks and Dylan F. Williams. *Characteristic Impedance Determination Using Propagation Constant Measurement.* IEEE Microwave and Guided Wave Letters 1.6 (1991), pp. 142–143.

[53] Roger B. Marks and Dylan F. Williams. *A general waveguide circuit theory.* Journal of Research of the National Institute of Standards and Technologies 97.5 (1992), pp. 533–561.

[54] Maury-Microwave. *Introduction to tuner-based measurement and characterization.* Application Note 5C-054. (2009), pp. 1–8.

[55] Maury-Microwave. *Vector-receiver load-pull measurement.* Application Note 5A - 051. (2011), pp. 1–8.

[56] S. R. Mazumder and P. D. van der Puije. *Two-signal method of measuring the large-signal S-parameter of transistors.* IEEE Transactions on Microwave Theory and Techniques 26.6 (1978), pp. 417–420.

[57] Juliana Müller and Bert Gyselinckx. *Comparison of active versus passive on-wafer load-pull characterization of microwave and mm-wave power devices.* IEEE Microwave Theory and Techniques International Symposium (MTT-IMS) Digest. San Diego, CA, USA. 24.9 (1994), pp. 1077–1080.

[58] Kaoru Narita and Taras Kushta. *An accurate experimental method for characterizing transmission lines embedded in multilayer printed circuit boards.* IEEE Transactions on advanced packaging 29 (2006), pp. 114–121.

[59] José C. Pedro, Luís C. Nunes, and Pedro M. Cabral. *A Simple method to estimate the output power and efficiency load-pull contours of class-B power amplifiers.* IEEE Transactions on Microwave Theory and Techniques 63.4 (2015), pp. 1239–1249.

[60] M.A. Pulido-Gaytan and J.A. Reynoso-Hernandez. *Sensitivity of FET parasitic elements extraction due to uncertainty on TRM calibration structures.* International Journal of RF and Microwave Computer-Aided Engineering. 31.12 (2021). DOI: 10.1002/mmce.22889.

[61] M.A. Pulido-Gaytan et al. *The Impact of Knowing the Impedance of the Lines Used in the TRL Calibration on the Load-Pull Characterization of Power Transistors.* 86th Automatic Radio Frequency Techniques Group (ARFTG) Conference Digest. Atlanta, GA, USA. (2015), pp. 5–9.

[62] Manuel Alejandro Pulido-Gaytan et al. *Vector network analyzer calibration using a line and two offset reflecting loads.* IEEE Transactions on Microwave Theory and Techniques 61.9 (2013), pp. 3417–3423.

[63] Manuel Alejandro Pulido-Gaytan et al. *LZZM: An extension of the theory of the LZZ calibration technique.* 84th Automatic Radio Frequency Techniques Group (ARFTG) Conference Digest. Boulder, CO, USA. (2014), pp. 5–9.

[64] Manuel Alejandro Pulido-Gaytan et al. *On the implementation of the LZZ calibration technique in the S-parameters measurement of devices mounted in test fixtures.* 85th Automatic Radio Frequency Techniques Group (ARFTG) Conference Digest. Phoenix, AR, USA. (2015), pp. 5–9.

[65] Manuel Alejandro Pulido-Gaytan et al. *Generalized theory of the Thru-Reflect-Match calibration technique.* IEEE Transactions on Microwave Theory and Techniques 63.5 (2015), pp. 1693–1699.

[66] Manuel Alejandro Pulido-Gaytán et al. *Calibration of a real-time load-pull system using the generalized theory of the TRM technique.* 88th ARFTG Microwave Measurement Conference Digest, San Francisco, CA. (2016), pp. 1–5.

[67] Manuel Alejandro Pulido-Gaytán et al. *Determination of the Line Characteristic Impedance Using Calibration Comparison.* 91st ARFTG Conference, 2018. Conference Digest (2018), pp. 1–5.

[68] Francesc Purroy and Lluís Pradell. *New theoretical analysis of the LRRM calibration technique for vector network analyzers.* IEEE Transactions on Instrumentation and Measurement 50.5 (2001), pp. 1307–1314.

[69] Jussi Rahola. *Power Waves and Conjugate Matching.* IEEE Transactions on Circuits and Systems II: Express Briefs 1.1 (2008), pp. 1205–1215.

[70] J. Apolinar Reynoso-Hernandez and Everardo Inzunza-Gonzalez. *Comparison of LRL (m), TRM, TRRM and TAR, calibration techniques using*

the straightforward de-embedding method. 59th ARFTG Conference, spring 2002. 32 (2002), pp. 93–98.

[71] J. Apolinar Reynoso-Hernandez et al. *What can the ABCD parameters tell us about the TRL?* 79th Automatic Radio Frequency Techniques Group Conference Digest. Montreal, QB, Canada. (2012), pp. 1–4.

[72] J.A. Reynoso-Hernandez et al. *Using lines of arbitrary impedance as standards on the TRL calibration technique.* 81st Automatic Radio Frequency Techniques Group (ARFTG) Conference Digest. Seattle, WA, USA. (2013), pp. 1–4.

[73] J.A. Reynoso-Hernandez et al. *Advances in Microwave Large-Signal Metrology: From Vector-Receiver Load-Pull to Vector Signal Network Analyzer and Time-Domain Load-Pull Implementations.* Electronics. 11.7 (2022), pp. 1–21.

[74] J. Apolinar Reynoso-Hernández. *Unified method for determining the complex propagation constant of reflecting and nonreflecting transmission lines.* IEEE Microwave and Wireless Components Letters 13.8 (2003), pp. 351–353.

[75] J. Apolinar Reynoso-Hernández et al. *Transmission Line Impedance Characterization Using an Uncalibrated Vector Network Analyzer.* IEEE Microwave and Wireless Components Letters 30 (2020), pp. 528–530.

[76] Patrick Roblin. *Nonlinear RF Circuits and Nonlinear Vector Network Analyzers: Interactive Measurement and Design Techniques.* New York, NY: Cambridge University Press, 2011. ISBN: 0890065055.

[77] Patrick Roblin et al. *New trends for the nonlinear measurement and modeling of high-power RF transistors and amplifiers with memory effects.* IEEE Transactions on Microwave Theory and Techniques 60.6 (2012), pp. 1964–1978.

[78] Rogers. *RO4000 Series High Frequency Circuit Materials.* Data Sheet (2015), pp. 1–4.

[79] David E. Root, Jan Verspecht, and Mhai Marcu Jason Horn. *X-Parameters: Characterization, Modeling, and Design of Nonlinear RF and Microwave Components.* Cambridge University Press, 2013. ISBN: 9780521193238.

[80] Andrej Rumiantsev and Nick Ridler. *VNA calibration.* IEEE microwave magazine 21.3 (2008), pp. 86–99.

[81] Randeep Saini et al. *An intelligence driven active load-pull system.* 75th Automatic Radio Frequency Techniques Group (ARFTG) Conference Digest. Anaheim, CA, USA. (2010), pp. 1–4.

[82] Michael B. Steer et al. *Comments on" An accurate measurement technique for line properties, junction effects, and dielectric and magnetic parameters[with reply].* IEEE Transactions on Microwave Theory and Techniques 40 (1992), pp. 410–411.

[83] Yoichiro Takayama. *A new load-pull characterization method for microwave power transistors.* IEEE Microwave Theory and Techniques International Symposium (MTT-IMS) Digest. Cherry Hill, NJ, USA. (1976), pp. 218–220.

[84] Keysight Technologies. *Keysight 85052C 3.5mm precision calibration kit, User's and Service Guide.www.keysight.com.* Application Note 5989-8575 (2019).

[85] Keysight Technologies. *Nonlinear Vector Network Analyzer (NVNA).* Application Note 5989-8575 (2019).

[86] Valeria Teppati and BColombo R. Bolognesi. *Evaluation and reduction of calibration residual uncertainty in load-pull measurements at millimeter-wave frequencies.* IEEE Transactions on Instrumentation and Measurement 61.3 (2012), pp. 817–822.

[87] Valeria Teppati and Andrea Ferrero. *A new class of nonuniform, broadband, nonsymmetrical rectangular coaxial-to-microstrip directional couplers for high power applications.* IEEE Microwave and Wireless Components Letters 13.4 (2003), pp. 152–154.

[88] Valeria Teppati et al. *Accuracy improvement of real-time load-pull measurements.* IEEE Instrumentation and Measurement Magazine 56.2 (2007), pp. 610–613.

[89] Valeria Teppati et al. *Microwave measurements part 3: Advanced nonlinear measurements.* IEEE Instrumentation and Measurement Magazine 11.6 (2008), pp. 17–22.

[90] Reydezel Torres-Torres et al. *Analytical characteristic impedance determination method for microstrip lines fabricated on printed circuit boards.* International Journal of RF and Microwave Computer-Aided Engineering. 19 (2009), pp. 60–68.

[91] Arturo Velazquez-Ventura and Antonio Lazaro. *Application of CAD load-pull techniques in mixer design.* Microwave and Optical Technology Letters 36.4 (2003), pp. 320–323.

[92] Scott A. Wartenberg. *RF measurements of die and packages.* Artech house., 2002. ISBN: 9781580532730.

[93] Dylan F. Williams and Roger B. Marks. *Transmission Line Capacitance Measurement.* IEEE Microwave and Guided Wave Letters 1.9 (1991), pp. 243–245.

[94] Dylan F. Williams and Roger B. Marks. *LRM probe-tip calibrations using nonideal standards*. IEEE Transactions on Microwave Theory and Techniques 43.2 (1995), pp. 466–469.

[95] Tudor Williams et al. *A digital, PXI-based active load-pull tuner to maximize throughput of a load-pull test bench*. 83rd Automatic Radio Frequency Techniques Group (ARFTG) Conference Digest. Tampa, FL, USA. (2014), pp. 1–4.

[96] John Wood. *Behavioral Modeling and Linearization of RF Power Amplifiers*. Norwood, MA: Artech House, 2014. ISBN: 9780521193238.

[97] José Eleazar Zuñiga-Juarez. "Técnicas de calibración LRL, LRM y LRRM para corregir los errores sistemáticos del analizador de redes vectorial y su impacto en la caracterización de transistores de altas frecuencias". PhD thesis. Centro de Investigación Científica y de Educación Superior de Ensenada, Baja California, Mexico., 2011.

[98] J. E Zúñiga-Juárez, J. Apolinar Reynoso-Hernández, and Andrés Zárate-de Landa. *A new method for determining the characteristic impedance Zc of transmission lines embedded in symmetrical transitions*. 1st ARFTG Microwave Measurement Conference, 2008. Conference Digest (2008), pp. 1–5.

[99] J. E Zúñiga-Juárez et al. *A new analytical method to calculate the characteristic impedance ZC of uniform transmission lines*. Computación y Sistemas. 16 (2012), pp. 277–285.

Index

About the Authors

J. Apolinar Reynoso-Hernández (Member, IEEE) was born in Tacupa Michoacán, Mexico. He received his degree in electronics and telecommunications engineering from ESIME-IPN, Zacatenco, Mexico in 1984, his M.Sc. degree in solid state physics from CINVESTAV-IPN, Zacatenco in 1985, and his D.E.A. and Ph.D. degrees in electronics from the LAAS, CNRS, Université Paul Sabatier, Toulouse, France in 1987 and 1989, respectively. In 1990, he joined the Department of the Electronics and Telecommunications, CICESE, Ensenada, B. C., Mexico, where he is currently a professor. His areas of specialized research interests include high frequency on-wafer measurements, high frequency device modeling, linear and non-linear active device characterization, and modeling. He led CICESE's Microwave Group to obtain the Best Interactive Forum Paper Award five times. Since 2013, he has been serving as the TPC for ARFTG and ARFTG-MTT Workshop Organizer 2019 and 2020. ExCom ARFTG member from 2021–2026. He has been Associate Editor of Transactions On Microwave Theory and Techniques since 2022 and a reviewer for many IEEE journals.

Dr. Manuel A. Pulido Gaytan was born in Mexico in 1986. He received his B.Sc. degree in Electronic Engineering from the Autonomous University of Zacatecas (UAZ), Zacatecas, México in 2010, and his M.Sc. and PhD. degrees from the Center for Scientific Research and Higher Education at Ensenada, Baja California (CICESE), Ensenada, México in 2012 and 2016, respectively. Since 2016, he has worked as RF Device Modeling and Characterization Engineer in different multinational organizations within the USA microwave semiconductor industry. As both graduate student and professional, Dr. Pulido-Gaytan's main activities have been related to the development of new techniques for characterizing and modeling microwave devices, as well as the design of low-noise and power amplifiers, where his work has contributed to the development of several products that have been successfully launched to production in the mobile phone industry.